Praise for The Agile Entrepreneurship Series

"Touting 'agile entrepreneurship,' a process he borrowed from software developers, Dontha cleverly structures the book with 30 chapters, each highlighting a discrete business-building area or task. Every chapter begins with the story of a different small-scale entrepreneur who employed the agile approach to succeed, providing welcome credibility."

—*Kirkus Reviews*

"It's all about working smarter, not harder. *The 60 Minute Startup* is the epitome of this approach: a blueprint for success, not to be missed.

—Midwest Book Review, D. Donovan, Senior Reviewer

"Thoughtful applications of resources make all the difference to new businesses. Follow Dontha's valuable information to avoid the pitfalls and watch your business grow!"

—San Francisco Book Review

"Even if you never planned to start a business online, Ramesh K Dontha makes you want to start one. The message in this book is so unmistakably clear that it helps readers find clarity with their own thoughts and understand the opportunities that are there for them. 5 stars out of 5 stars."

—Readers' Favorite, Christian Sia

"*The 60 Minute Startup* is written in a confident voice and in a conversational style that engages readers, a style that reflects the author's business experience and acumen. It is filled with the business advice that any new entrepreneurs need. A highly recommended read for business success. 5 stars out of 5 stars."

—Readers' Favorite, Romuald Dzemo

"Based on agile methodology opposed to rigid implementation, this self-help guide to establishing a business gives hope to entrepreneurs with conservative funds and limited time to get a promising venture started."

—The BookLife Prize

THE
60-MINUTE
PODCAST
STARTUP

THE 60-MINUTE PODCAST STARTUP

A PROVEN SYSTEM TO START YOUR PODCAST IN 1 HOUR A DAY AND GET YOUR INITIAL AUDIENCE IN 30 DAYS (OR LESS)

RAMESH DONTHA

The Agile Entrepreneurship

Folsom, CA

The Agile Entrepreneurship
Folsom, CA
www.RameshDontha.com
Send feedback to Contact@the60minutestartup.com

Publisher's Cataloging-In-Publication Data

    Names: Dontha, Ramesh, author.
    Title: The 60-minute podcast startup : a proven system to start your podcast in 1 hour a day and get your initial audience in 30 days (or less) / Ramesh Dontha.
    Description: Folsom, California : The Agile Entrepreneurship, [2021] | Series: [The Agile Entrepreneurship series] ; [volume 3] | Includes bibliographical references.
    Identifiers: ISBN 9781733465175 (hardcover) | ISBN 9781733465182 (softcover) | ISBN 9781733465199 (ebook)
    Subjects: LCSH: New business enterprises. | Podcasts--Economic aspects. | Entrepreneurship. | Agile project management. | Success in business.
    Classification: LCC HD62.5 .D663 2021 (print) | LCC HD62.5 (ebook) | DDC 658.11--dc23

Special discounts for bulk sales are available.
Please contact Contact@the60minutestartup.com.

To my wife Sunanda and to our daughters Megha and Nidhi. You have always motivated me by saying that my books will have at least three readers.

Contents

What's Next?

www.The60MinuteStartup.com

Start your podcast today the agile way! As a valued reader, you get all the templates, scripts, and tools you need to produce your 60-minute podcast, all at no cost to you.

Want support from other entrepreneurs on this thirty-day journey? You can also join the free private group to ask questions, bounce around ideas, and even find your first listeners. Go to the link below.

To get started on your entrepreneurial journey about starting a business or starting your own podcast, Ramesh Dontha provides courses and coaching. Find out more at www.The60MinuteStartup.com.

www.The60MinuteStartup.com

Tell Me What You Think

Let other readers know what you thought of *The 60-Minute Podcast Startup*. Please write an honest review for this book on Amazon or on your favorite online bookshop.

★ ★ ★ ★ ★

Tune In

Several entrepreneurs featured in this book, in *The 60-Minute Startup*, and in *The 60-Minute Tech Startup* appear on The Agile Entrepreneurship Podcast with author Ramesh Dontha. Listen at www.RameshDontha.com/Podcast.

The Secret to Launching a Successful Podcast with the Knowledge and Skills You Already Have

Jordan and Blake both want to start a podcast.

Jordan is a professional marketer but believes business interviews are the only way to monetize a podcast. He copies successful business podcasts he listens to without understanding why those podcasts work. Meanwhile, Blake is a nutritionist and loves her field of expertise, so she decides to start a health podcast to embrace her existing knowledge base.

Jordan decides he has to get the biggest guests, the best equipment, and the perfect specific theme right away, or no one will want to listen to his show. He buys the most expensive recording software with complicated features and vows to learn every function before he starts recording. Blake interviews people she already knows in her field, uses simple recording equipment she has access to even if it's used or open source, and covers generic topics based on what she already knows. She sticks to what's simple and familiar to help her start faster.

Jordan only wants superstars on his podcast because he assumes no one will want to listen to his show otherwise. But he's never built any connections in entrepreneur circles and keeps getting turned down by the strangers he invites. When his unrealistic expectations backfire, he gets frustrated and doubles down. Meanwhile, Blake invites health-care professionals she already knows from her existing practice. She tests as many ways to promote her podcast as possible to see what works. When she experiences a software setback or gets turned down by a guest, she's not devastated because she expects that she will learn as she goes instead of demanding perfection from the start.

Jordan is desperate to make money on his podcast but sticks solely to traditional methods of monetization, like sponsorship. Blake thinks outside the advertising box. She sees her podcasts as a content-creation system to find new ways to attract clients, create big opportunities, and promote her products.

Within thirty days, Jordan launches his podcast, but it's ill prepared and ill defined. He has one interview done, but no other pitches worked, so he has no other guests. He's still spending so much time trying to find guests and figure out the expensive enterprise software that he hasn't spent much time on promotion. His first episode gets only two dozen downloads, which crushes his hopes of launching into immediate success. Meanwhile, Blake has eight interviews ready, one per week for two months, to be released on the same day. Her first episode gets two dozen downloads, and she's looking forward to growing that number over time with better promotion methods. She continues exploring new tactics to book and interview more guests and get more downloads.

Jordan is still struggling with production and can't even think about consistent promotion, much less monetization. He's confused about his end goal, stressed about making money, and has no idea how to create value for himself or his audience. It's been one month, and he's already thinking about giving up. But Blake is focused on production, promotion, and proactive monetization. She knows exactly what her podcast is for and is actively proceeding toward her end goal. She's confident about making money and prepared to do this for at least six months.

The path Jordan took to start his podcast is the one most aspiring podcasters assume they have to follow. It's painstakingly slow, creates a ton of stress, and makes no guarantees. It's high risk, with the hope of high reward based on concepts that worked for other podcasters, but more often than not, the show fizzles out. Sooner or later, podcasters like Jordan get disillusioned by their low listener counts and decide no one wants to hear them speak.

Podcasters like Blake go a different way, the path of lowest risk with the highest upside based on what works for them personally. The only objectives and key results (OKRs) Blake cares about in the beginning revolve around building an audience and providing value through existing means. So while Jordan's podcast crashes and burns on launch, Blake's catapults her ahead.

Blake has a system built on doing what works in incremental steps to reach success. That's why Blake's show will be around for a long time. So can yours.

What Agile Software Development Can Teach Us about Successful Podcasts

If you've ever worked in the technology industry, you've no doubt heard of agile software development. It's a project management methodology that emphasizes speed, flexibility, and results over perfection, precision, and tradition. But you may not be familiar with the origin story of the agile method—or how agile translates into the fastest way to start a tech company and get paying customers without pursuing outside funding or quitting your day job.

In February 2001, seventeen software developers from around the world met for a three-day retreat to get drunk over problems with their bosses and share ideas about how to get projects done on time.[1] This was a casual event. Little planning went into it beyond booking the lodge. Yet what emerged from their conversations was the Agile Software Development Manifesto. This profound document changed the way developers plan projects, collaborate with coworkers, report

---

1   "History: The Agile Manifesto," Agile Manifesto, accessed April 7, 2020, www. agilemanifesto.org/history.html.

to management, and meet deadlines. The Manifesto states four primary values that should guide every project:

1. Individuals and interactions over processes and tools.
2. Working software over comprehensive documentation.
3. Customer collaboration over contract negotiation.
4. Responding to change over following a plan.[2]

The agile approach revolutionized software, made Silicon Valley the tech capital of the world, and made possible the Airbnbs, Dropboxes, and Instagrams we can't imagine living without. If agile software founders can build companies worth billions in a garage, imagine what the agile approach can do for you! Well, you don't have to imagine—*you* can be the next Blake. Agile makes it possible.

Let's back up a second and see how *Merriam-Webster* defines the word *agile*.

1. marked by the ready ability to move with quick and easy grace.
2. having a quick, resourceful, and adaptable character.[3]

Typical podcasters invest years and fortunes building their podcasts, which may never produce a profit. Agile podcasters are different. They believe that *good enough to make money is good enough to make money.* Agile podcasters choose what works for them over what has worked for others. Fast over methodical. Done now over done perfectly. Authentic downloads over vanity metrics such as likes.

The agile method is so superior to every other path to starting any new project that I dedicated my first book to launching a successful business: *The 60-Minute Startup: A Proven System to Start Your Business in 1 Hour a Day and Get Your First Paying Customers in 30 Days (or Less)*. The book teaches how to inventory your interests, passions,

---

2   Kate Eby, "Comprehensive Guide to the Agile Manifesto," Smartsheet, July 29, 2016, www.smartsheet.com/comprehensive-guide-values-principles-agile-manifesto.

3   *Merriam-Webster*, s.v. "agile," (accessed April 7, 2020), www.merriam-webster.com/dictionary/agile.

and business ideas and turn them into a viable business that attracts paying customers by the end of month one.

The book you're reading now—*The 60-Minute Podcast*—marks the third installment of the 60-Minute Startup series. This book addresses the unique challenges and also opportunities of podcasting as opposed to general business success advice. What it takes to build a thriving podcast without giving up looks very different from starting up a creative endeavor in a totally different industry.

If you're not looking to start a podcast, I highly recommend that you also read the second book in this series, *The 60-Minute Tech Startup*. So we're clear on what is in scope for the book you're reading now, let me define what I mean by podcast. I like Wikipedia's definition because it's so broad.

> A podcast is an episodic series of spoken word digital audio files that a user can download to a personal device for easy listening. Streaming applications and podcasting services provide a convenient and integrated way to manage a personal consumption queue across many podcast sources and playback devices.4

In other words, a podcast can be an iTunes audio download, a live YouTube interview show, videos recorded on your smartphone or computer without notes or preparation, and so on.

I'm passionate about helping you turn your podcast idea into a viable revenue-producing show because I've done that for myself with *The Agile Entrepreneurship Podcast*, which reached the top 500 iTunes entrepreneurship podcasts; *The Data Transformers*, which is a top-250 technology podcast globally on iTunes; and *AI—The Future of Business*, a top-ranking web series on BrightTALK. And I can tell you that running your own show is tremendously fun, fulfills your creative side, and allows you networking opportunities you can't even imagine right now.

I'll share a bit about myself first so you know where I'm coming from. I was trained as a mechanical engineer and an industrial engineer (but never worked for even a single day as either). I started my career

---

4   *Wikipedia, The Free Encyclopedia*, s.v. "podcast," (accessed January 26, 2021), www.en.wikipedia.org/wiki/Podcast.

as a systems analyst and traveled in India, Europe, and the US. After a while, I realized that my interests were elsewhere, so I left that cushy job and got my MBA. I then worked in marketing, management, business development, and strategic planning areas for Fortune 100 companies, enjoyed the work, learned (and earned) a lot, and traveled the world all over again.

Throughout all these years of paying my dues to the corporate world, I dreamed of being an entrepreneur. I started my first business, a tutoring business, just after my undergrad college while waiting to figure out what I wanted to do with my life. That venture didn't last long. Since then, I have started four different companies, gone through the ups and downs of being an entrepreneur, sold two of them, and continued to run my remaining ventures. I currently actively run a strategic management consulting business focusing on data analytics. You can get more information on that at www.DigitalTransformationPro.com. I also run a successful podcast you can find at www.RameshDontha.com/Pod-cast.

Based on my experience building my popular startup podcast, *The Agile Entrepreneurship Podcast*, I figure that you probably have a lot of questions about starting your own podcast, questions such as:

- Where do I start?
- How do I pick a concept?
- How do I find guests?
- How do I pick the right recording equipment and software to start?

You'll find the answers to these and other questions in the pages ahead. For now, let's tackle the first concern: where to start.

Agile planning encourages you to work on something small, execute it quickly, get feedback, evaluate what's working and what's not, and adapt your plan from there. This process of small, fast, and repeated cycles is known as "iterative." So to get you started, we're going to approach podcasting with an agile framework. That means we're going to start small, based on what you have available. I'll show you exactly how to do that in the next few chapters.

Unfortunately, many people who start a podcast end up like Jordan. They take too much time to launch and build their podcast because they're so focused on getting everything perfect right from the start. My recommendation is to have a strict timeline to start and launch your podcast so you do not get stuck in the perennial loop of learning.

In *The 60-Minute Startup*, I introduced the fastest way to start your business based on the agile method. Set aside sixty minutes a day for thirty days to work on your business, and you should be able to build an initial listener base within that time.

Because you're launching a podcast in your spare time, your timeline will look a little different. You'll use the thirty days to test your new podcast concept, get feedback, validate the podcast idea, and get the minimal equipment you need to start recording. Again, you'll apply agile principles to do this with less stress, with less money, and in less time.

How to Use The 60-Minute Podcast

60 minutes a day x 30 days = 1 viable podcast

That's my promise to you if you read and apply everything in the pages to come. As I wrote this book, I skipped all the typical (bad) podcasting advice and trimmed essential tasks down to the *most* essential. Ever heard of the 80/20 rule? Eighty percent of the results you want come from 20 percent of your effort. Well, *The 60-Minute Podcast*, like its predecessor *The 60-Minute Startup*, is more like a 99/1 rule. That means you're doing one thing here and one thing there that have a big impact. I can tell you from experience that what *feels* productive often isn't. The only important activities to start and build a podcast aren't the ones that take days and weeks to finish. In fact, they don't take as long as you might expect. Only a few tasks lead to a growing listener base. For aspiring podcasters like you, that is the only goal: listeners. No listeners, no monetization.

This book is meant to be read one chapter a day for thirty consecutive days. In each chapter, I'll show you what to do next and how to get it done fast. Each chapter opens with a story of a successful podcaster who took the agile approach to that day's activities. That way, you can see for yourself what that task looks like when it's done right. You'll also see a checklist of steps for the day, suggested time for each, and any templates you need to get the day's work done in sixty minutes or fewer. If you read one chapter every day and complete the tasks, you will most likely have a growing listener base in thirty days. For any podcasting tasks that change over time (e.g., my recommended recording software), I'll direct you to www.The60MinuteStartup.com for my most up-to-date resources. That way, you can start your own podcast today, whether you're reading this book in 2021, 2031, or beyond!

Over the next month together, we're going to borrow principles from the Agile Software Development Manifesto and apply them to your podcasting dream. We'll do so using something called a *scrum*. Our scrum is the recipe we'll follow that tells us what essential ingredients we need for our agile framework. Agile is the *what*; *scrum* is the how. Let's break down the critical elements of our scrum:

- Sprint
- Sprint team
- Product backlog
- Sprint backlog
- Story points
- Sprint planning
- Daily standup
- Sprint review
- Sprint retro

In a scrum, you **sprint** to get meaningful work accomplished over a defined time period. In *The 60 Minute Startup*, we'll have two fifteen-day sprints, so we'll finish in thirty days exactly.

A **sprint team** is composed of the owner, the scrum master (a servant-leader who manages and coaches the team while tracking deliverables), and anyone else working on the scrum. In our case, the owner is you, the scrum master is me, and the rest of the team is anyone who helps you accomplish the day's task or the week's sprint (a lawyer, a copywriter, etc.). You're taking a team approach in everything you do. You're not in this alone. In every chapter, you're going to witness how another agile podcaster accomplished the same tasks you're going to do that day. You're essentially joining a sprint team of superstars. That way you'll start and finish each day feeling motivated to keep going.

The **product backlog** is simply a list of the requirements needed to develop a meaningful product. In our case, the product backlog is the entire list of all the tasks you'll accomplish in this book over the next thirty days to start your podcast and develop your audience. You're prioritizing the important things. Your only objective right now is to get your first listeners. Prioritize every task that gets you closer to that goal.

When you start a podcast, figuring out what to do and in what order is one of the hardest tasks. That's why I've defined what the highest-priority actions are and laid them out for you in the order you should do them.

The **sprint backlog** is the list of things that need to be accomplished during each sprint. In our case, the first fifteen-day sprint covers your existing field of expertise, your specific podcasting format, your proof of concept, selecting your recording software, and everything else you need to start working with guests and gathering listeners, including contact templates.

The second fifteen-day sprint is all about attracting and engaging with potential guests and listeners, approaching them with the opportunity for an interview, and leveraging those relationships to generate even more interviews so you can build a profitable, sustainable guest list. Congratulations! You now have a real podcast.

Story points are essentially a breakdown of each task into steps with how long it takes to do each step. Don't worry about being exact. An estimate is good enough. Knowing what you know now, what kind of turnaround can you expect on this step? You make an initial estimate of how much effort you'll need to put into a task, then you keep updating that estimate based on how things are going. Every day, I estimate that the tasks I give you in each chapter will take sixty minutes. If one day's tasks go more quickly, feel free to take on the next chapter and dive in if you have the time. If another day's tasks end up taking you two hours, extend that chapter's work into the next day. It all evens out in the end. Flexibility is part of the agile approach.

When you're **sprint planning**, you're figuring out what input needs to go into a particular sprint. Don't worry—I've already made these decisions for you. Your sixty-minute task list every day for the next thirty days is done and waiting for you.

The **daily standup** is a self-assessment. Every day during a software project, team members keep each other accountable by asking each other short questions like "What did you get done today? How are you doing since yesterday? What do you need to carry over to work on tomorrow?" At the end of each day (chapter), you'll see this simple two-choice standup, and you'll check the appropriate box. You'll also answer two additional questions about your experience that day.

Did you complete today's tasks?

❑ Yes
❑ No

If no, what do you need to carry over to work on tomorrow?

_____

What did you learn about your business (or yourself) today that will serve you in the future?

_____

Will you need to check the first or third box often? It's doubtful. Your task list for each day includes only the essential activities that lead to your first, second, and third (and so on) paying customers.

A **sprint review** is exactly what it sounds like. At the end of each week, you review what you've accomplished. As I did in *The 60-Minute Startup*, this book will give you the opportunity to review what you got done during each sprint and to prepare for what's next.

In a **sprint retro**, you ask yourself how you can improve. What did you learn during the last sprint that you can apply in the future? I provide a space for you to answer this question after each week for your sprint retros.

I have a unique perspective on the agile approach because I used it as a product manager, a consultant, and an entrepreneur. Before agile methodology, software development was tedious. We got detailed project requirements from the client in months one and two. We didn't start developing the product until the third month. Yet everybody knew the requirements would change. So when change occurred—priorities shifted, key personnel quit, the economy crashed, et cetera—the first two months were wasted. When we took the agile approach, we went straight to prototyping in the first month. I'd show our client the first prototype of their user interface and logo, and they'd either say, "Yes, perfect," or "No, that's not what I meant." If they didn't like it, we went back to the drawing board. Either way, we lost no time.

This approach applies to all entrepreneurial journeys, including podcasting. When you draw up a business plan, it's never perfect, right? It's going to change because change happens. Not everything you write down on day one is going to stick. The same is true when designing a podcast. So you keep evolving that business plan, or podcasting plan in this case. If you have a general idea about what you're going to do, that's good enough. Go do it, see what happens, and adjust your next steps.

By the end of this book, you will have concrete steps you can put into action to monetize your podcasting idea and start moving toward them right away. All this is made possible by the agility approach. Applying the agility concept to podcasting means we will focus on:

- Concept agility
- Format agility (guest, monologue, reviews, tutorials)
- Guest selection agility
- Equipment agility
- Podcast prep (record, edit, distribute) agility
  - Budget agility (podcasting is one industry in which people can start with almost zero budget)
- Promotional hustle
- Monetization hustle

Starting a podcast may seem daunting, but if you apply the agile method to building your audience, you'll be amazed at what you can accomplish in just sixty minutes a day.

This Book Will Work for You If . . .

You pick a podcasting format and topic that are right for you. It's better to stick with what you already know than to do what others are doing just because it worked for them. In a coming chapter, we'll discover the type of format and topic that's right for you. For now, I'd like to give you a preview of the many podcasting formats and topics that could work for you and your situation.

You're a Self-Help and Business Podcaster

Self-help podcasts are vital to listeners looking to grow themselves into a stronger, wealthier, more resilient person. If you're helping people learn to run a business, you might run a podcast about marketing strategies. Podcasts focused on teaching people to manage their physical or mental health could make use of personal success stories from guests who've overcome major health issues. Or maybe you want to teach your listeners how to cultivate new skills to start a side hustle. All these self-help areas have enormous potential to encourage and educate listeners and draw a huge following according to the need you address.

You're a Religious Podcaster

As the world turns to digital realities and society becomes more isolated, the search for meaning has become all the more urgent. You might run a podcast educating people about your personal religion, like Christianity or Buddhism, and how to apply that religion's principles to a modernized life. You could discuss the impact of the global world on the living structure of specific organized religions, or take things in the other direction and show listeners how to build nonorganized spirituality into their day with gratitude and appreciation for the universal divine.

You're a Current Events and Entertainment Podcaster

Modern events move at the speed of light. The in-vogue celebrities of today are tomorrow's public enemy number one. Entertainment shifts as technologies change and demands adapt. For example, before the 2020 pandemic, audiobooks were growing as a popular medium with so many consumers driving long commutes to their offices, but the pandemic threw a wrench in those gears by shifting much of the population to working from home. You could discuss not only current entertainments but also the methods of delivering those entertainments and help your audience stay abreast of the latest developments.

You're a Society and Culture Podcaster

We live in an increasingly global society, which gives you infinite inspiration to draw from when you discuss culture. What's happening on one side of the globe has become immediately relevant on the opposite side, so you'll never be hurting for topics to discuss. And helping your audience stay up to date on culture changes across the globe positions you as the expert in analyzing how to shift business or social approaches to better suit various settings.

If your podcasting idea relates in any way to the podcast types I've listed, *The 60-Minute Podcast* is for you. Because the agile approach is for you. To this day, I follow agile in every new entrepreneurial venture I start. I prototype, prove it works, tweak what doesn't, and evolve in the direction of getting customers. I'm going to teach you how to do the same, whether you're starting your first podcast or your next one.

Really, Ramesh? All that in thirty days? you're probably thinking. Yes, success is not only possible; it's *likely* when you build your podcast the agile way. In the coming pages, I'm going to introduce you to other agile podcasters who started their podcasts in record time. While aspiring nonagile podcasters sat around brainstorming show names, these entrepreneurs booked guests and gathered listeners. Don't waste another minute on articles, emails, podcasts, tutorials, or lengthy business books that get you trapped in busywork. If you want to launch your podcast and make money now, the agile way is the only way for you.

"But Ramesh, What If . . . ?"

With a promise like mine (a growing listener base headed toward monetization in one hour a day in just thirty days), you might still have your doubts. I would if I were you. I bet you're thinking, *This agile podcasting thing seems all well and good, but what if it doesn't work for me? I don't have any podcasting experience.*

Fair question. The entire point of *The 60-Minute Podcast* is that you don't *need* any podcasting experience. All the experience you need— knowing what to do and when to do it—comes with this book. You don't have to figure out anything by yourself. You're following in the footsteps of successful podcasters who've built thriving podcasts the agile way. So the question isn't "Will this work for me?" It's "Can I follow instructions?" If you can, you have nothing to worry about. This book can and will work for you.

But Ramesh, what if I don't have a following, I'm on a shoestring budget, and I don't have much time? Won't it take like a month to pick a topic and get comfortable with my voice? Not with *The 60-Minute Podcast*, it won't. Every day, I will give you what you need to get your tasks done. For example, you don't have to be an expert on your specific podcasting topic, because you're interviewing experts in that industry. All you need to know are the questions that you should ask them. When it's time to write your emails to reach out to guests, I'll give you my templates. I've already tested these templates, so all you have to do is copy, paste, modify as necessary, and send. The same goes with every other complex task. Even drawing up show introductions. The hard work is already done for you. If you can download a template, you can start your own podcast and gather listeners in thirty days or fewer.

Still worried you're the exception? That the agile approach I teach may not work in your industry, with your idea, or for you personally? *The 60-Minute Podcast* is for serious entrepreneurs who want a real podcast with real listeners as soon as possible. Now, not every podcasting model allows that. For example, if you're starting a podcast focused on millionaire entrepreneur interviews, you need networking skills to find guests, time to send out invitations for the show, and time for your guests to find an opening in their schedules. You can expect many months before you sign up your first millionaire guest. I could not pass the red-face test and promise you a booming guest list in thirty days. But if your dream is to open an interview podcast, why not start with smaller guests you can reach now? You won't need a lot of upfront commitment to start your agile business. After you use this book to get your paying listeners in thirty days, then you can work on attracting your millionaire guests. First things first. It's the agile way.

No matter your industry, idea, or personality, all you need to get started is this book, a computer, and an internet connection. A dedicated space such as a home office or a comfortable garage helps, but it's optional. If the laptop lifestyle appeals to you, you can build your own podcast from anywhere.

Now, before we begin day one, I have a bit of trivia for you.

Did you know there are 500,000 active podcasts in the world?[5] That's a lot. Most have hardly any listeners. A few get millions of downloads a week.

My goal is not to help you throw yours into the mix. It's to guide you to launch one that stands out from all the others on your subject matter. You may not have an award-winning podcast one month from now. However, you can expect to be one of the few who starts, builds, and grows a viable, profitable podcast.

Let's begin.

---

5   "2021 Podcast Stats & Facts (New Research From Jan 2021)," Podcast Insights, January 1, 2021, www.podcastinsights.com/podcast-statistics.

Day 1: Decide Why You're Doing This

Your Podcast Goals

It All Starts with a (Broad) Idea

Now that you've decided to launch a podcast, consider your proposed podcast goal. What do you want to accomplish with your podcast? Too few podcasters answer this question before they begin. If you don't know what "success" as a podcast host looks like, how will you know this is worth it? You won't.

Having a goal helps you persist through the hard segments of your own podcasting journey. And there will be hard times, just as there are with any new venture. To help you get started, let's take a look at goals that successful podcasters had when they started.

- To generate content that can be repurposed into other media
- To generate leads for an existing (or new) business
- To be recognized as a leader in a specific industry
- To share an important message you are passionate about
- To have fun engaging with other people and sharing knowledge
- To raise awareness about specific causes
- To attract revenue-generating opportunities

- To teach (and learn from) others
- To entertain the audience
- To motivate and inspire

Do you have to pick just one? No. All are available to you, and possible. And can you change your goals once you've set them? Of course you can. Maybe your very first goal was to have fun interviewing interesting people, but your podcast got more traction—more downloads—than you expected. As a result, you may decide to adjust your goal to build a business around the podcast. Still, it is essential that you start with at least *one* goal as you are more than likely to flail in uncertainty without one.

The initial goal for my first podcast, *Agile Entrepreneurship*, was to generate content I could share on my blog and later in a book, which I did. That's how *The 60-Minute Startup* came to be. The podcast's success also helped me share an important message—that aspiring entrepreneurs can overcome inertia to get real, paying customers as quickly as possible. So I continued the podcast even after I accomplished my first goals. My third goal has been to establish myself as a recognized industry expert.

What are other podcast hosts saying?

"I started our podcast with a goal of producing meaningful business, marketing, and entrepreneurial content to help others."
—Jack Fleming, Host, *Process Over Profits*

"Our *Media Futurists* podcast goal was to have nominees in our Story & Technology festival and New Media Film Festival be able to share different aspects of what they do, in a format outside of the normal live, in person Q & A after their screening, press releases, social media posts. The podcast also would allow some of our team members to put their creative input into this new podcast idea."
—Susan Johnston, Host, *The Media Futurists Podcast*

Make Your Podcast Goal a SMART One

As with any goals, it is better to have SMART goals. So what is a SMART goal? The acronym SMART stands for **specific**, **measurable**, **achievable**, **relevant**, and **time bound**.

The goals I mentioned in the beginning of the chapter are not SMART goals, but they are a starting point. Your activity for today will be to start with your own personal goal for your podcast and make it a SMART goal. Let me give you one example.

My starting point for my first podcast was "to generate content that can be repurposed into other media."

I made this into a SMART goal: "I want to launch a podcast about starting an online business by interviewing at least four entrepreneurs within thirty days."

This goal is:

- Specific: Podcast about starting an online business.
- Measurable: Four entrepreneurs.
- Achievable: Interviewing four entrepreneurs in thirty days is achievable.
- Relevant: Starting a business is relevant to me because I started businesses in the past.
- Time bound: Launch the podcast in thirty days.

Other SMART goal examples include:

- I want to grow my audience to ten thousand downloads per month within three months.
- I want to make $1,000 per month from my podcast within six months.
- I want to be the number one podcast on iTunes for my topic in four months.

How do you know you'll achieve your goal? Well, you don't. But you will be able to measure your progress toward your goal on a regular basis (I suggest daily monitoring) and take necessary steps to adjust your actions as necessary.

Your Podcast Goals: Now It's Your Turn

Check off the box beside each task as you complete it.

❑ 1. Write down your goal: 15 minutes.

On a piece of paper, write down your podcast goal. For this first step, don't worry about making your goal a SMART goal.

❑ 2. Make your goal a SMART goal: 30 minutes.

Now make this goal a SMART podcast goal. And here are some questions to guide you.

Specific: What do you want to accomplish? Why is this a goal? Who is this for? (We'll dig into specific audiences on Day 2, but think about your broad category for now.) What general topic/theme do you want to target? (We'll dig into topic/theme in detail in a later chapter.)

Measurable: What metrics will help you measure your progress toward your goal?

Achievable: Do you have the skills to achieve this goal? Or can you obtain them? Is the timeframe reasonable? Are the metrics realistic?

Relevant: Is the topic/theme relevant for the audience? Is the topic/theme relevant for you?

Time bound: What is the deadline, and is it realistic?

❑ 3. Review your SMART goal with a friend/family member/colleague: 15 minutes.

So far, you have set your SMART goal by yourself. Do a simple sanity check by reviewing it with a close friend or a trusted family member. Do they think it makes sense?

Daily Standup

Did you complete today's tasks?

❑ Yes
❑ No

If no, what do you need to carry over to work on tomorrow?

What did you learn about your business (or yourself) today that will serve you in the future?

Day 2: Choose Your Target Listeners

Listener Avatars and Personas

L et me first start with a disclaimer. You can either pick a podcast theme or topic first, then decide on the audience that may be interested in that topic, or you can start with your target audience and then decide on the topics they may be interested in. Both approaches are OK. Day 3 is about selecting a podcast topic/theme. You can switch Day 2 and Day 3 if you feel comfortable selecting your theme first. For reference, I selected my audience first and then narrowed down the podcast topic for all my podcast series.

Why Do You Need an Avatar?

Ask any product designer or developer how they came up with their product. I'm certain they'd mention a customer avatar or persona they had in mind. They are attempting to address their potential customer's problems or needs and hoping that their product will meet their needs.

Your podcast is your product. You want your product to address their needs. But don't you want to know what your future listener does for a living, their age range, their challenges in life, and where they hang out? Wouldn't this information help you target your podcast to address their needs as opposed to some random attempt? Of course it would.

To be as specific as possible, focus on at least the following bare minimum.

- What they do for a living (e.g., small business owners)
- Age range (e.g., between thirty and fifty years)
- Gender (e.g., 60 percent men and 40 percent women)
- Where they live (e.g., mostly suburbs in the US)
- What their primary challenges are (e.g., access to financing to grow their businesses)
- Their likes (e.g., involved in kids' activities)
- Other activities (e.g., local chamber activities)

While I was coming up with my listener avatar, I was advised to come up with a name like Accountant Adam, Coach Christie, et cetera. I preferred to come up with a persona with broader characteristics, rather than a specific avatar. It is your choice.

> "We started our podcast as a way to help women who are going through a career transition due to layoffs, COVID-19, or other reasons find their way back into the workforce."
>
> —Mary Sullivan, Host, *Sweet but Fearless*

> "Although we have younger listeners, we target an audience who is 62 or older. It's my experience that this is the audience least likely to listen to podcasts, and in 2016, they didn't even know what a podcast was! Of course that is changing, and we have noticed an uptick in our downloads."
>
> —Kathe Kline, Host, *Rock Your Retirement*

How to Create Your Avatar

How do you know who your podcast is for? Simple. Here are some ideas for you to do research on coming up with your ideal listeners.

- If you have a business, interview your clients.
- If you have an email list, send a survey.
- Use software like SurveyMonkey to survey real people.
- If you have a website, look at your visitor analytics to see what people are reading most on your site. You can find visitor demographic information as well, such as country of origin.
- Research similar podcasts; specifically, read the reviews to see what listeners want that you can deliver (or do better).
- Read industry blogs and forums for popular threads and articles. What people already like, they will like from you.
- Copy someone else's persona. Research industry influencers to find out which audience they are targeting.

I have designed an ideal listener avatar template to help you in this regard. Please download the template from https://rameshdontha.com/members-options/.

While completing the template, focus on the problems you can solve for them and the opportunities you can identify for them. This will help you on Day 3 when you'll select your podcast theme and topic.

Your Podcast Audience: Now It's Your Turn

Check off the box beside each task as you complete it.

- ❏ 1. Start with the template and complete as much as you can: 20 minutes.

First, download the ideal listener avatar template. Given that you have been thinking about a podcast, you have a basic idea of who you want to target. Complete the template as much as you can. Don't worry about making it perfect.

❏ 2. Do additional research: 30 minutes.

Next, reach out to your additional research sources. These could be industry blogs/forums, existing podcasts in your genre, or sending out surveys/emails to your customers. If you need to send out emails/surveys, you need to wait until you receive their responses and then come back to complete the template.

❏ 3. Review your listener avatar with a friend/colleague/ mentor/coach: 10 minutes.

Throughout the book, I'll be including this review section where I believe it is important. Try to have a friend/colleague/coach who can give a second opinion, just to make sure that you are not too far out in left field.

Daily Standup

Did you complete today's tasks?

❏ Yes
❏ No

If no, what do you need to carry over to work on tomorrow?

_____

What did you learn about your business (or yourself) today that will serve you in the future?

_____

Day 3: Coming Up with a Podcast Theme and Topic

What Is Your Podcast About?

What You Need to Know about Podcast Genres

Podcasts come in all different themes and focus on a variety of topics. As of October 2020, the top 10 podcast genres are:

1. Comedy
2. News
3. True crime
4. Sports
5. Health and fitness
6. Religion and faith
7. Politics
8. Self-help and productivity
9. Investigative journalism
10. Finances

This bar graph illustrates and compares the popularity of these and other top podcast genres.[6]

Leading podcast genres in the United States in October 2020

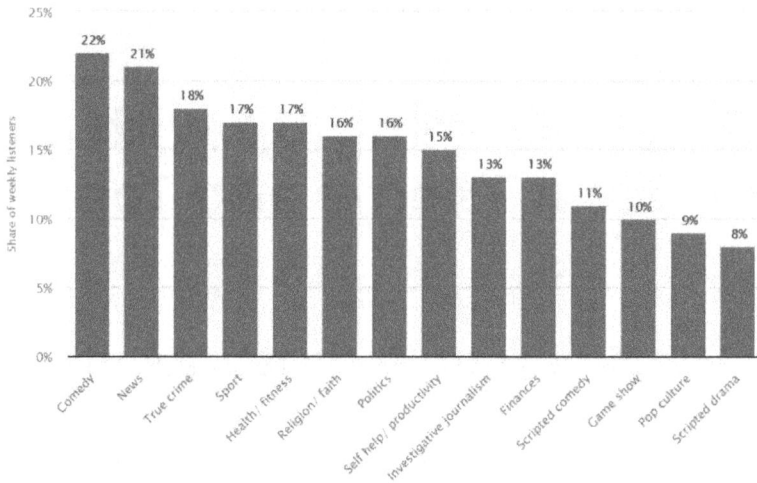

© Statista 2021

Right now, you probably have one of two questions—or both.

1. Should my podcast be in one of the top genres?
2. If I do choose a top genre for my podcast, does that mean I'll face tougher competition?

My answer to the first question is "Probably yes." But if you have a burning desire and passionate interest for a non–top ten genre like scripted comedy, game show, pop culture, or scripted drama, go for it.

The answer to the second question is "Competition is always good." The top ten genres are a safe place to start because those listeners already exist. *Your* listeners already exist. And they're consuming

6   Statista Research Department. "Leading podcast genres in the United States in October 2020," Statista, March 4, 2021, www.statista.com/statistics/786938/top-podcast-genres.

28

content right now that you can supplement via your own show. Unless you have a compelling reason to pursue a less-listened-to niche, stick with what works.

> "Everyone was launching a podcast! We decided, even though our particular after-show television space—and space related to *The Golden Girls* as a show and fandom in particular—was crowded, to carve a very particular niche of 'scholarship' out of our conversation. We spent time collecting academic papers on television scholarship, contacted and interviewed professors who had written papers on *The Golden Girls,* and ensured we were writing SEO articles filled to the brim with exciting new fandom angles that would appeal to people who might not even be searching for a podcast to listen to."
> —Sarah Royal, Host, *Enough Wicker: Intellectualizing The Golden Girls*

> "*The Media Futurists* podcast is about 'honoring stories worth telling.' New media is an infinite catalyst for story & technology. We chose to create themes around several episodes. The Creators Series, for example. We discovered this led people to more easily figure out what they wanted to listen to and why."
> —Susan Johnston, Host, *The Media Futurists*

Podcast Topic Ideas

Now that you have a general lay of the podcast genre land, let's look at specific podcast topics. Here are four questions to ask yourself to help select a topic.

- **What subject are you interested in?** The plan is for you to be in the podcasting business for a while, so it goes without saying that you should be interested in and curious to learn about your topic.

- **What are your strengths?** Play to them. Pick either a topic you know well or one for which you have access to subject matter experts who can share their wisdom. Depending on which show format you select later on (e.g., narrative, guest interview, etc.), identify your strengths (or your ideal guests') so your topic becomes easy-peasy for you to record episodes soon.
- **How will you be a unique show host?** Will you focus on a specific audience that has not been addressed yet—niche down? Or do you have a unique storytelling talent?
- **Will you be able to monetize your podcast?** At some point, you want to monetize your podcast, whether it is through sponsorships, sell up, referrals, or some other means. So select a topic that is monetizable. Of course, picking a topic in the popular genres mentioned above is a good start.

Now let's talk about specific topics. The following ideas will get you started. To download a comprehensive list, head on over to https://rameshdontha.com/members-options/.

- Talk about **people and places**. Your podcast could be about soccer moms, for example, or hairdressers, plumbers, travel on pennies, or hidden vacation places.
- Talk about **your own journey.** If you have a unique journey with ups and downs, that story could be a great topic. My only caution is to make it interesting so your audience doesn't get bored. Combine it with others' journeys like yours.
- Talk about **podcasting** (seriously). I've met several YouTubers who make money teaching others how to start their own YouTube channel. So why not teach podcasting? Interview other podcasters for additional advice.
- Talk about **learning a new hobby or skill**. Podcast about a hobby or skill you yourself are learning. Your audience may want to learn along with you.

- Talk about and with **your ideal customers**. Let's say you're a digital marketer serving lawyers. Why don't you invite your customers and prospects to talk about their profession? Only good things can happen.

Check www.RameshDontha.com for an exhaustive list.

Your Podcast Theme and Topic: Now It's Your Turn

Check off the box beside each task as you complete it.

❏ 1. Write down your podcast theme/genre: 15 minutes.

On a piece of paper, write down your top three podcast themes. Under each of the themes, write down pros and cons for each. Why you want to focus on that theme and why you would not focus on that theme. This exercise will help bring more clarity. Then pick the best theme. Remember that if you decide to change your theme later on, you'll have this sheet to come back to pick your next-best choice.

❏ 2. Select your top three podcast topics: 30 minutes.

Go through the above short list or the longer list from our website and short-list your top three topics.
Ask yourself these four questions to help you short-list.

- What is your sweet spot? What topics are you most comfortable with?
- Who is my audience? Go to Day 2 and put yourself in your audience's shoes.
- What topics can set you apart?
- How will you monetize your podcasts?

❏ 3. Review your theme and topics with a friend/family member/colleague: 15 minutes.

As you've figured out by now, I want you to keep getting feedback on your selections just to make sure that your choices are reasonable from the perspective of the people who know you best.

Daily Standup

Did you complete today's tasks?

❑ Yes
❑ No

If no, what do you need to carry over to work on tomorrow?

_____

What did you learn about your podcast (or yourself) today that will serve you in the future?

_____

Day 4: Your Name, Keywords, and Cohost

The Smart Way to Rank Well

How to Be Descriptive, Catchy, and Memorable

What's in a name? Well, a lot. I'll let iTunes tell you why.

Pay close attention to the title, author, and description tags at the and level of your podcast. Apple Podcasts uses title, author, and description fields for search. The metadata for your podcast, along with your podcast artwork, is your product packaging and can affect whether your podcast shows up in relevant searches, and how likely users are to subscribe to it.

Make your title specific. A podcast named Our Community Bulletin is too vague to attract many subscribers, no matter how compelling the content.[7]

From why to how next, here are my guidelines for naming your podcast.

---

7 "A Podcaster's Guide to RSS," Apple, accessed June 6, 2021, https://help.apple.com/itc/podcasts_connect/#/itc2b3780e76.

- **Make your name relate to your topic.** A podcast about starting a business or being an entrepreneur probably should have either *business* or *entrepreneur* in the name so the audience knows exactly what the podcast is about.
- **Go broader than your niche.** This may be counterintuitive, but let me give an example. If you are an expert in hiking and you named your podcast *Hiking*, what happens when you want to cover other outdoor activities as your show gains an audience?
- **Be clever but not too cute.** This is a tough one. The challenge in being too cutesy is that the audience may not get it right away, and search engines may not know what search terms to rank for based on name alone. So be judicious.
- **Use at least one keyword in your name.** As you may know, keywords are what the audience searches by for what they are looking for. As an example, my first podcast is about starting a business, so I included the word *entrepreneurship* in the name: *The Agile Entrepreneurship Podcast*. My second podcast was about data analytics, so I named it *Data Transformers*.

If you're stuck on a name, browse the top 100 podcasts on iTunes or Spotify for inspiration. There are a few resources listed in the Your Turn section below to get you started. While you're at it, don't forget to secure a domain name that is either the same as your podcast or close to it.

"I created the *Fading Memories* podcast because I wanted to help other families cope with the daunting task of caring for Alzheimer's or dementia patients. Mom had had Alzheimer's for more than 15 years, and I felt I had a lot to share. I quickly learned that I also had more to learn. Pivoting to an interview-style show, I now share stories from other caregivers who have walked this same path. The goal of my podcast is to be a voice of knowledge and inspiration. Hearing these stories at any moment when those are most needed is why I chose the podcast platform."

—Jennifer Fink, Host, *Fading Memories*

"My podcast is unscripted with me talking about movies I watch or things I want you to watch. Occasionally, I slip in things I deal with in real life or what I am working on with my career goals. My most popular episodes so far are when I talk in lengthier episodes about things in the news, such as how I began reviewing Harry Potter films for my website, and suddenly J. K. Rowling started on her anti-trans comments, requiring me to ask if I need to ignore the creator, focusing on the art."

—Nicole Russin-McFarland, Host, *CinematNIC*

Use Keywords to Rank

Podcast directories look at your podcast descriptions to help listeners discover podcasts. To help them in that process, use keywords judiciously in your podcast descriptions. Make your description informative but also discoverable by search engines. As the keywords will be very handy not just in podcast description but also in naming your podcast episodes later on, prepare a list of a large number of relevant keywords that you may want to tap into later on.

Here are some tips for keyword usage.

- **Use your audience's language.** If you are already expert in your domain, hopefully you'll know some keywords they use.
- **Research using free keyword tools.** Use Google Keyword Planner (a free keyword tool) to research what the audience is searching for and the search volume. Even though the search tool is not about podcasts, you'll get an idea of what your audience is searching for in general.
- **Use a combination of long-tail keywords and shorter keywords.** An example of a long-tail keyword may be "starting a business with no money in California" or "how can I get

a job in data science in one year?" Long-tail keywords will help you rank quickly in that specific area.

- **Check out other podcasts in your domain to give you ideas.** Obviously, it goes without saying not to copy your competition, but get some ideas on how they became successful or what keywords they are targeting.

Consider a Cohost

Even though there is no perfect place to discuss this particular topic, this chapter is probably the latest that you can go to if you are considering a cohost. After this, it becomes very difficult to bring in a cohost as you may have already decided on the topic/theme and name.

There are pros and cons about a cohost. If you can find a cohost you can work with, the podcast won't be as overwhelming. Depending on each other's expertise, you can share the day-to-day operations of running a podcast, which can be overwhelming.

Name Your Podcast: Now It's Your Turn

Check off the box beside each task as you complete it.

❏ 1. Short-list your podcast names: 20 minutes.

Prepare a short list with priorities of your podcast names. Here are some tools to help you with synonyms and related words.

- OneLook: This is a powerful dictionary tool that lets you search by partial word or topic.
- RhymeZone: This website will let you search for rhyming words, homophones, similar-sounding words, phrases, and famous lyrics or poems.
- Fantasy Name Generators: As the name says, it helps you find fantasy names if you are trying to be clever.

- Instant Domain Search: This website will help you with all unused domain names with all kinds of extensions like .dog or .cat if you can't find your .com domain.

❑ 2. Prepare a list of keywords: 30 minutes.

Now prepare a list of short-tail as well as long-tail keywords related to your topic.
Here are some free keyword research tools.

- Google Keyword Planner
- Google Trends
- Ahrefs Keyword Generator
- Keyword Sheeter

❑ 3. Review your names with a friend/family member/colleague: 10 minutes.

Send your short-listed podcast names to a mentor or a close friend and get their feedback.

Daily Standup

Did you complete today's tasks?

❑ Yes
❑ No

If no, what do you need to carry over to work on tomorrow?

_____

What did you learn about your business (or yourself) today that will serve you in the future?

_____

Day 5: Format, Length, Schedule

Put Your Personal Stamp on Podcasting

Show Format

Today is all about the logistics of your podcast and its episodes. This is a sandbox—there is no right or wrong podcast format. There is only what works for you. So you know what those options are, let's look at the most popular formats.

- **Interview**: By far, this is the most popular podcast format. Book interesting guests and ask them questions they haven't heard before. You can have a single host or multiple hosts interviewing the guests. All my podcasts are interview style.
- **Monologue**: Solo podcasts extend your brand and authority. The challenge is the feeling that you are talking to yourself. If you think you can make it interesting with a monologue, go for it.
- **Cohosts**: Discussion and banter between two hosts work for the right people. This, of course, does away with the

"talking to yourself" issue and can make it interesting without any guests.

- **Miscellaneous**: You can have a multiguest roundtable, a scripted documentary format, or some other combination of the above.

It's all about what you are comfortable with. Pick one and move on.

> "After building a large following on TikTok, there were many followers who told me that they would listen to my podcast if I had one. I thought I'd like to do it, but it wasn't until another podcaster had me on that I realized how much meaningful content could be delivered to followers (45 min episodes as opposed to 1 min videos). Going in depth and having listeners stay engaged for so long to build a stronger connection."
>
> —Dr. Kim Chronister, Host, *Love to Heal with Dr. Kim*

Episode Length

The length of your podcast goes hand in hand with the schedule of your podcast. If you are planning to publish an episode daily, it may not be a great idea to have an hour-long episode daily unless you have really engaging content that can demand your listener's attention for sixty minutes a day.

Research has shown that the average podcast length is about forty-three minutes per episode with a median of thirty-eight minutes.[8] Another statistic to keep in mind is that the average commute time in the US is around twenty-five to thirty minutes.[9] I try to keep my episodes

---

8   Dan Misener. "I analyzed 10 million podcast episodes to find the average length," Pacific Content, October 25, 2018, https://blog.pacific-content.com/how-long-is-the-average-podcast-episode-81cd5f8dff47.

9   "United States | Mean travel time to work(minutes), workers age 16 years+," United States Cenesus Bureau, accessed June 6, 2021, www.census.gov/search-re-

around twenty-five in general, though I have had episodes as long as forty-five minutes.

In general, try to target fifteen minutes if you have a daily show, sixty minutes for a weekly podcast, and maybe ninety minutes for a monthly podcast. Pick your length based on your schedule.

Publishing Schedule

When do you publish episodes? The answer depends entirely on your ability to record, edit, and prepare the episodes. Podcasting can be an all-consuming job. If you have an interview-style podcast, you need to schedule the guests, prepare the guests, and record the episodes.

The successful podcasts have one thing in common in addition to great content. They have a consistent schedule. So whether it is one episode per week or daily, pick a schedule and be consistent. Ideally, you should aim for at least one episode per week so the listeners are not left hanging for too long.

Format Your Podcast: Now It's Your Turn

Check off the box beside each task as you complete it.

❏ 1. Select your podcast format: 30 minutes.

So what is it going to be? Brainstorm on different formats. For an interview style, do you have the ability to attract interesting (and may be influential) guests? For a monologue, have you noticed that people were hanging on to your every word when you were speaking at an office function? Or do you prefer a scripted fiction/nonfiction type? Write down the pros and cons of a couple of formats and make the final decision.

---

sults.html?q=commute&page=1&stateGeo=none&searchtype=web&cssp=SERP.

THE 60-MINUTE PODCAST STARTUP

❏ 2. Think about length and schedule: 20 minutes.

As I mentioned earlier, the length and the schedule go hand in hand. What is your bandwidth? Can you consistently commit to a daily podcast of your selected format? Or a weekly or a monthly podcast? To some extent, this choice also needs to be looked at in the context of your podcast SMART goal. Just follow the guideline of fifteen minutes for a daily podcast, sixty minutes for a weekly podcast, or ninety minutes for a monthly podcast.

❏ 3. Review your choices with a friend/family member/colleague: 10 minutes.

Send your short-listed podcast format, length, and schedule choices to a mentor or a close friend and get their feedback.

Daily Standup

Did you complete today's tasks?

❏ Yes
❏ No

If no, what do you need to carry over to work on tomorrow?

What did you learn about your business (or yourself) today that will serve you in the future?

Day 6: The Right Recording Equipment

Audio Gear to Fit Any Budget

The All-Important Microphone

Podcasting has an extremely low barrier to entry. All you need is a mobile phone or a computer, a podcast-hosting platform (some are free), and boom! You're in business. If you have compelling content, listeners will even be willing to put up with inferior audio quality.

Is that the goal, though? Publish the lowest-quality content you can get away with? Absolutely not. We're here to deliver the most optimal quality podcast that fits your budget. If you have a few dollars to spare, the best investment you can make is a USB microphone. It's a recording studio that plugs right into your desktop or laptop computer. This one investment goes a long way toward establishing you as a professional.

Here are my recommended recording mics as of this writing, but you can find an updated list at https://rameshdontha.com/members-options/.

Entry-Level Price (< $100)

- Audio Technica ATR2100x
- Samson Q2U
- Rode Smartlav+

Premium Price (> $100)

- Rode Podcaster/Procaster
- Blue Yeti
- MXL 990
- Shure SM58
- AKG Lyra

"I use the ATR2100 microphone."
—Tanya Fox, Host, *Fox Talks Business*

"I have a ATR2100 microphone."
—Stephen Warley, Host, *Life Skills That Matter*

"I use a Rode Podcaster microphone."
—Dave Rael, Host, *Developer on Fire*

Nice-to-Have Equipment

If your budget allows, consider other accessories such as a boom arm or shock mount to hold your microphone steady, headphones for playback clarity, and a pop filter to soften hard sounds. These are my favorite brands and products.

- Rode PSA Boom Arm
- On Stage MY-420 Studio Microphone Shock Mount
- Audio-Technica ATH-M20X Professional Studio Monitor Headphones
- Neewer Professional Microphone Pop Filter

Upping Your Game

If you want the best possible podcast quality money can buy, get yourself a professional-level microphone, high-end headphones, and a mobile digital voice recorder whose one job is making your voice sound the best it possibly can from wherever you happen to be recording.

High-End Mics and Headphones (> $250)

- Heil Sound PR 40
- Focusrite Scarlett 2i2 (2nd Gen)
- Audio-Technica ATH-M50x Headphones

If you know you'll always record your podcasts from your desk or studio, the equipment listed above is fine. But what if you want to record at a conference or on-site or in a remote location? You'll need a mobile digital voice recorder. Here are my top picks.

- Zoom H5/Zoom H6
- Zoom PodTrak P4
- Rode Rodecaster Pro

Buying That Gear: Now It's Your Turn

Check off the box beside each task as you complete it.

❑ 1. Finalize your budget and recording gear: 40 minutes.

Do some additional research on microphones, headphones, and other gear. Come up with a budget and finalize your recording gear choices.

❑ 2. Buy your equipment: 20 minutes.

Make your purchases and make sure that you'll get the equipment a day or two before your first recording.

Daily Standup

Did you complete today's tasks?

❏ Yes
❏ No

If no, what do you need to carry over to work on tomorrow?

_____

What did you learn about your business (or yourself) today that will serve you in the future?

_____

Day 7: Podcast Software

The Only Technology in Podcasting

Audio-Only Recording and Editing?

Podcasts. They're audio only. Right? Well, yes and no. As far back as podcasting has existed as a medium, there have always been audio-only shows. With the growing worldwide popularity of and access to high-speed internet, an increasing number of hosts are adding video. Some are releasing audio-only podcasts on iTunes while concurrently publishing the full video on a video sharing site like YouTube. We'll cover the advantages of video later on, but first let's focus on audio only. It doesn't matter what people see or don't see if they can't get past what they hear.

Depending on your budget and technical ability, you have multiple options for audio-only recording and editing.

Audacity (free): This is a free, open-source software with a steep learning curve. I use Audacity but limit my edits to the bare minimum. Only the obvious mistakes like technical difficulties do I cut from episodes. Luckily, there are a ton of tutorials and lessons to make it easy. Go to https://rameshdontha.com/members-options/ for Audacity podcast editing tutorials.

GarageBand (free): For Apple users, GarageBand is the ideal choice. It is free on Mac and iPhones and has a more intuitive user interface than Audacity.

Adobe Audition (paid): Adobe Audition is an excellent choice for professionals looking for best-in-class audio editing capabilities. This software has a much steeper learning curve but is rich with features.

Alitu (paid): Alitu is another paid software—basically Adobe Audition but with an easier-to-figure-out user interface. It also has editing features and episode-building tools.

Most aspiring podcast hosts will do nicely with one of these options.

"Podcasts aren't rocket science. However, there are literally one hundred different little things you should do right from the outset that will make the difference between your podcast sounding professional or like you're producing it out of your garage. Since I didn't know what I didn't know when starting out, we hired a podcast consultant who helped me understand what those one hundred things were."

—Guy Nadivi, Host, *Intelligent Automation*

"We used to do a mixed in-studio and remote set-up (via Skype) but have moved to a completely remote set-up (Skype or WhatsApp, depending on the host configuration). We host files on our own server but can be listened to across most platforms. We use a variety of recording equipment depending on the hosts: in my own studio I've got a mixer, SM58 microphones, an external backup voice recorder, headphones, headphone amp, and a computer running Audacity."

—Jenny Mathiasson, Host, *The C Word: The Conservators' Podcast*

Audio and Video Recording and Editing

Midway through the first season of *The Agile Entrepreneurship Podcast*, I decided to record video interviews in addition to audio. Video recordings provided me with the added visibility of YouTube in addition to the usual podcast directories. Video expanded my show's reach with minimal effort. Here is the software I recommend if you're going to record video and audio together.

Zoom (free and paid): Zoom has become the go-to virtual meeting platform. Their video recording provides above-average audio quality as well. For recording shorter episodes, it's free (forty-five minutes at the time of this publication). After you record with Zoom, you may need additional software for editing the video. Zoom also saves the audio recording separately if you just want to use the audio.

Skype (free): Even though Skype has better name recognition than Zoom and a longer track record, many hosts have stopped recording Skype because of poor quality and a cumbersome setup for recording. That said, if you already use Skype, it may make sense for you to go with what you know.

SquadCast (paid): SquadCast is my favorite premium software for video recording. It records audio with separate tracks for each speaker, which makes it super easy to edit later on.

Which video recording app should you choose? Well, which do you already use? Better to get right to implementation than waste any undue energy making a decision. Do what works. Set it and forget it.

Video Editing

After recording the video, you may want to edit the video to cut out unwanted portions or stitch various pieces together. There are lots of video editing tools out there, but the best tool I'll recommend is Camtasia, a paid software that is relatively easy to learn. It also has a lot of options to add text and animation and extract only audio. For a complete tutorial on how I edit my videocasts, log in to the member portion of https://rameshdontha.com/members-options/.

Select Your Software and Try It Out: Now It's Your Turn

Check off the box beside each task as you complete it.

❑ 1. Select your recording software: 20 minutes.

These are the questions you want to ask yourself to help decide on the software.

- Are you a solo host, or do you have a cohost?
- Is your podcast an interview format or monologue?
- Are you recording audio only or video/audio?
- How tech savvy are you?
- Do you want to add in special effects or special music elements?
- Do you want free software, or are you willing to pay for the software?

Based on the answers to these questions, the information provided above should lead you to a recommendation.

For example, Audacity is the best option for free, solo-host, audio-only, monologue type of podcast. It may have some learning curve, but it is manageable.

❑ 2. Test your software: 40 minutes.

Once you have selected your software, you can start testing the software right away as almost all of them have a free trial period even if they are paid. If you are testing multiple software products, forty minutes may not be sufficient. Budget your time accordingly.

Daily Standup

Did you complete today's tasks?

❑ Yes
❑ No

If no, what do you need to carry over to work on tomorrow?

What did you learn about your business (or yourself) today that will serve you in the future?

Day 8: Episode Research and Prep

Upfront Planning Saves Hours

Whether your podcast is an interview, monologue, or show with your cohost, a checklist will save you countless hours, smooth out your operations, make you more productive, and reduce stress.

So what should be on the checklist? Here's mine for inspiration.

❏ **Guest-related items**
  ❏ An invite letter/email to guests. The letter should mention your podcast theme, audience, details about the recording (audio or video), prompting them for proper microphone/camera setup.
  ❏ Guest headshot, bio, social media account info.
  ❏ Guest introductions for the episode.
  ❏ Guest background information (to be filled out by the guests).

I typically put together a Google form with all guest-related items and send my podcast guests that link to fill out the form prior to the recording.

❏ **Checklist for the day of the recording**
  - ❏ Camera/microphone setup.
  - ❏ Make sure all apps in the background are disabled. I reset my laptop before the recording just to be safe.
  - ❏ Guest introduction is ready.
  - ❏ Shared notes (cohost or guest) or your own research notes and question prompts.
  - ❏ Specific call to action (your own or guest related).
  - ❏ Reminder to *press* that record button! So many times, hosts forget to press that button until midway through the recording. I have done it, too . . .

For a downloadable checklist and forms you can customize and use right away, visit https://rameshdontha.com/members-options/.

> "The biggest challenge with launching my podcast was making sure I had enough content ideas to produce a weekly show. If you take a look at what the show has accomplished so far, you'll see 100+ episodes that are all 20 minutes or more. I have very few guests, which means nearly all the content came directly from me. I wanted to make sure I had a viable concept, so I sat down and brainstormed fifty show ideas and topics before I ever recorded my first episode."
> —James Pollard, Host, *The Financial Advisor Marketing Podcast*

> "I did a podcast (Episode 20–The 5 Reasons a Podcast Fails) focused on the struggles of doing a podcast. It's hard to continue to put out original content and schedule it on a regular frequency. It's so much of having the right equipment and time. The timepiece is important so you can keep the creative juices flowing."
> —Nathan Webster, Host, *Let's Talk Marketing*

Scripting Your Episodes

How you script your episodes depends on your personality and expertise. Podcasts are more like conversations, whether it is with your guests or your audience. Hence, I am not a big fan of word-for-word scripting of an episode. Still, I'll mention various approaches so you can pick your own type.

Word-for-word podcast script: This may be suitable for hosts who are just getting started and not very confident about managing the flow. The downside to scripting the entire episode is that it may sound mechanical and monotonous unless you are good at hiding that you are reading a script.

Detailed episode plan: Instead of a word-for-word script, you'll jot down your entire plan, including your introductions, flow, and closing. It is midway between a word-for-word script and a simple bullet list.

Simple bullet list: This is my preference. I put together the most salient points I want to address about the topic or the guest. I may not follow the flow of my list, but this list is more for making sure that I don't skip them. My episodes are conversational, and I let the flow dictate how I am progressing from point to point. That's why this format is my favorite.

Whatever format you choose, I highly advise you to have a pre-interview with the guest before the actual episode. Even though this is optional, it is highly desirable. I suggest scheduling a fifteen-minute pre-meeting to discuss where you and the guest can find value, discuss the flow, and discuss any promotional items to offer to increase value to the audience. It is also another opportunity to remind the guests about your theme, requirements, et cetera.

Planning Tools

Tools to help you plan and get ready for your episodes will save you a lot of headache and time. Here are my favorites.

- **Calendar**: Calendly, Acuity
- **Forms and checklists**: Google Forms (my favorite), Typeform, JotForm
- **Guest data and recording backup**: Google Drive, Dropbox, OneDrive

I am sure there are a ton more tools for every possible task, but I try to keep my life simple.

Prepare For Your First Recording: Now It's Your Turn

Check off the box beside each task as you complete it.

❑ 1. Select your tools: 10 minutes.

If you are familiar with the tools mentioned above or you have a system already, you can skip this task. Otherwise, select your tools from the list above.

❑ 2. Prepare your checklists: 40 minutes.

Either copy the checklists form https://rameshdontha.com/members-options/ or prepare your own. Make sure to use some kind of cloud backup to store your documents.

❑ 3. Select your scripting format: 10 minutes.

Take the time today to research and review different scripting formats mentioned above. You don't need to script any episodes today; just be prepared with a format.

Daily Standup

Did you complete today's tasks?

❏ Yes
❏ No

If no, what do you need to carry over to work on tomorrow?

What did you learn about your business (or yourself) today that will serve you in the future?

Day 9: Guests Who Make (or Break) Your Podcast

Be My Guest

Finding the Right Guests

Today is relevant for you if you want to interview guests on your podcast right from the start or want to at least consider interviewing others for your show in the future. First things first. I'll state the absolute obvious: finding your first podcast guests can be quite challenging. Here are six tips based on my experience that will help ease you into it. I wish I'd known all these when I started *The Agile Entrepreneurship Podcast*.

1. Start with your friends and connections. You don't need the most popular guests to get started. You want some supporting guests initially to get the "podcast blues" out of the way.
2. Reach out to your existing customers (if you have any) or your ideal prospects. Everyone would love to share their expertise and their story, especially if it's free publicity.
3. Search other podcasts in your category on iTunes, Spotify, or some other directory. Send a note to either the hosts or their guests. They know about the power of podcasting already.

4. Search blogs or social media in your industry and reach out to the authors. I usually find great, compelling guests on LinkedIn, for example.
5. Attend a few conferences or get a list of speakers at these conferences. The speakers will turn out to be great guests.
6. Sign up with a podcast guest service. There are plenty of guest services like PodMatch.com, MatchMaker.fm, Kitcaster.com, Guest.Market, et cetera.

These are just a few ideas to build your initial guest list.

> "One of the struggles was getting into the rhythm of what to actually talk about each week—so I created a document that is just a running tab of ideas or things that my clients say or stuff I notice coming up in the media. I make a quick note so I always have a whole list of topics to choose from and then just start writing."
> —Danielle Savory, Host, *It's My Pleasure*

> "My final challenge was finding guests. I never imagined this would be difficult. As an ex-journalist, I know how keen people are to get onto TV, podcasts, magazines. But it seems journalists themselves don't care about the limelight! So I had to downgrade my ambition of once a week podcast production to once a month. However, once I got a few big names, it got easier. I've now had editors on from ITV News, *Huff Post*, *Red Magazine* and *Business Insider*, so it's easier to attract guests—but be prepared to be snubbed in the beginning!"
> —Helen Croydon, Host, *The Media Insider*

The Art of Interviewing

OK, you've found some guests. What are you going to talk to them about? Will you have enough questions to ask and topics to discuss? Here are six additional tips—these will get your guests talking. They require a little up-front time investment from you but will pay handsomely later in the form of a podcast interview people can enjoy.

1. Do research on their social media accounts or their blogs. My initial list of topics is based on what they have already been talking or writing about.
2. Have a pre-interview if possible. This doesn't have to be more than fifteen minutes. In the pre-interview, you'll get a sense of the guest's communication style. Are they talkative? Or do you need to "force" them to talk? You can also clarify topics of interest to them and topics they don't want to talk about.
3. Keep the conversation going. Don't spend too much time on their background intro. Get into conversation quickly.
4. Don't interrupt (too much). I usually let the guests finish their thoughts, but there are times when a guest keeps going, so I do interrupt by asking a clarifying question. This will break the monotony.
5. Have a set of core questions ready. These could be about their education, specific experience, products or service, or even personal interests. Some guests want the questions to be sent in advance.
6. Listen to your past interviews and other master interviews. Regularly listen to your own podcasts. Also spend time, especially initially, listening to other master interviewers on radio, TV, or other popular podcasts.

Managing Guest Lineup

The hard part is done . . . so you think. Some people get stuck in organizing the guest list and managing the schedules. I have a system—I use a master Google Sheet to manage my guest list. I list information such as their contact info, social media info, when I sent the initial invite, current status, and recording schedule time.

Some people use a calendar scheduling tool like Calendly to ask guests to pick a time. I am a little bit flexible and ask the guest to give three possible times within a range. And I schedule the most convenient time for both of us.

And don't forget to remind the guests a day or so before about the upcoming recording and what you need from them on the day of the recording. Also remind them about sitting in a quiet place and having an external microphone and other essential equipment.

That's all you need for now. Once the conversation gets going, allow yourself to enjoy the conversation. If you're a human being, you've already had plenty of practice talking. Which you are. And you do!

Get Ready with Guests: Now It's Your Turn

Check off the box beside each task as you complete it.

❑ 1. Prepare your initial guest list: 20 minutes.

Using the tips mentioned above, prepare your initial guest list. Depending on how much you've thought about this, this step may not need the twenty minutes.

❑ 2. Prepare your core interview topics and questions: 20 minutes.

Prepare your core set of questions you want to ask all your interviewees. Try to have a minimum of about ten questions. You may not ask all these questions, but at least you have a set to rely on.

❑ 3. Send your initial invites: 20 minutes.

Draft your invite email and send to your prospective interviewees to get the ball rolling.

Daily Standup

Did you complete today's tasks?

❑ Yes
❑ No

If no, what do you need to carry over to work on tomorrow?

_____

What did you learn about your business (or yourself) today that will serve you in the future?

_____

Day 10: Preparing to Record Your First Episode

Prepare, Practice, Perfect

Find Your Perfect Place

Today is all about getting ready to record your first episode. You'll not record today but prepare for it. As the old saying goes, if you have an hour to chop down a tree, spend the first fifty minutes sharpening the ax.

One of the most important decisions you'll make as you launch your podcast is recording location. If you're among the tiny percentage of hosts with a recording studio, you're in luck. I'm not. I record all audio and video in my home office. You don't need a professional sound studio either. A peaceful and quiet place will do.

If you are also planning to record video, make sure to have enough lighting on your face. (We've discussed specific equipment choices already.) I sit in front of a window with plenty of daylight. For night recording or cloudy days, I use a ring light.

For audio, make sure there are no audio distractions. Silencing any phones and devices is a must. Be conscious of other audio distractions like street noise, musical instruments in the house, or a chiming clock.

"I was fortunate enough to have a friend already podcasting with a local podcast network. After one visit, I decided it was a better use of my time and finances. For a reasonable fee, I get an hour of studio time, with an existing set-up that adds professionalism, as opposed to my living room or home office, an engineer. The studio provides automated services such as uploading and distribution to all audio podcast platforms, giving me the video portion for YouTube."

—Aalia Lanius, Host, *UNSUGARCOATED with Aalia*

"Getting started and figuring out what to do was the first big hurdle. I overcame that by getting started and showing up every week. Next challenge was creating content every week. Again, showing up every week made the difference. When you hold yourself accountable to showing up every week, you learn how to create new episodes. Another challenge was getting guests. By constant outreach, I started to get guests that eventually seemed out of reach, like the director of the National Park Service and the director of US Fish and Wildlife."

—Jody Maberry, Host, *The Park Leaders Show*

Set Up Your Recording Hardware and Software

By now, you should have, at minimum, an external microphone, speakers, and an external HD webcam. Other optional equipment such as a pop filter for your microphone and a microphone stand can be helpful as they improve sound quality and stabilize your environment, respectively.

You'll also need your recording software ready. As I mentioned earlier, I use Zoom software for recording both audio and video. Whatever is your preferred software, install it and test it. Play your test recording to work out the kinks. And now you are all set.

Review Your First Episode Outline

If you have the first guest lined up, use this time to do additional research on the guest. If your podcast is not an interview style, then use this time to do additional research about your initial topic.

Then prepare an outline. Whip up those core questions you have prepared in the prior chapter to prepare this outline. Practice your introduction. Practice your guest's introduction. Practice your mock episode with a friend or a family member to get rid of those butterflies. Practice, practice, practice before your first recording. Perfect practice makes perfect.

Prepare and Practice: Now It's Your Turn

Check off the box beside each task as you complete it.

❏ 1. Prepare your recording place: 20 minutes.

Pick a place and prepare the area.

❏ 2. Set up hardware and software: 30 minutes.

Set up your laptop and test out the setup.

❏ 3. Review episode outline: 10 minutes.

Practice your outline.

THE 60-MINUTE PODCAST STARTUP

Daily Standup

Did you complete today's tasks?

❏ Yes
❏ No

If no, what do you need to carry over to work on tomorrow?

_____

What did you learn about your business (or yourself) today that will serve you in the future?

_____

Day 11: Recording Your Very First Episode

When the Rubber Meets the Road

Introductions

Today is the day the proverbial rubber meets the podcasting road. If you've followed all prior steps, today should be a breeze. Otherwise, this could be a day when you'll think about quitting podcasting for good. Many people do give up because they didn't have discipline up until this point, but you don't have to be that person.

A common struggle podcasters have as they're about to record their first episodes is how to introduce themselves and the show. Tomorrow we'll go over intros and outros, which are standard introduction segments for all podcast episodes. Each episode could have its own unique intro as well.

One of the simplest introductions is to just say what the topic of the day is and get straight to it. If it's an interview podcast, introduce the guest and get going; that's what I do. Some hosts tease the audience about what's coming up, and they talk about anything interesting happening like a giveaway at the end. The hosts do this to keep the audience all the way till the end. I am fine with that strategy but try to balance it out. Research shows that the majority of the audience will tune out within a few minutes. So if you keep dragging in the beginning, you may lose your audience before you even start.

It's all about what you are comfortable with. Pick one and move on.

> "Our initial challenge was not having the proper equipment and recording directly from our phone. One thing about recording a podcast is making sure you have quality sound. We learned immediately that it could not be compromised and invested in booking time at a recording studio so that we would have access to their equipment. As we continued to record episodes and learn of ways to improve, we would implement them."
> —Ayanna Dutton/Delaila Catalino, Hosts, *Non-Corporate Girls*

> "I live on a relatively busy street, my neighbors have wind chimes, and you'd be surprised how raucous our water heater and refrigerator can get. Recording with so much background noise can certainly get frustrating. I don't have the money to invest in a home studio, so I've tried to make up for it with pillows, recording at nighttime, and clever editing in GarageBand."
> Winston Chang, Host, *Heard About*

Hit That Record Button

Now that you've planned your outline and intros, jump in by hitting that record button. It almost seems like common sense but remembering to hit that record button should be your most important task today. There are numerous examples of hosts forgetting to hit record until much later into the segment or, worse, even after the supposed "recording." Imagine you got your most important guest after a long time and, you forgot to hit the record button.

Keep talking. Don't worry about pauses, errors, *ums*, *ahs*, stutters, or repeats. All of them are editable. Make it informal and let the conversation rip. There are a few suggestions I'd make about microphone placement. Keep your microphone two to four inches away from your mouth. Most of the external microphones are good at picking up your audio from this distance. If you have a pop filter, that's even better as it softens your hard syllables like *p*'s.

Once you're done with the episode, try to conclude with a summary or thanking your guest as opposed to abruptly ending. This soft conclusion will prepare your listener that the episode is ending. And at the end, press that stop button. This is not as fatal as forgetting to hit the record button but still important.

Post-Recording Steps

Once you've completed your first recording, congratulate yourself. You've done it! But don't run away as some important stuff still needs to be done. First, rename your recording file with the date and the guest name or some name so you can remember.

Next, back up your recording, preferably on the cloud like Google Drive. (I use Microsoft OneDrive.) The backup is necessary because it saves you a lot of stress just in case your laptop crashes. Secondly, you can edit that recording from anywhere or deliver it to your outsourcer if you have outsourced editing.

Lastly, make important notes about the recording. I use a Google Doc, and I jot down the working title, guest name, and any key aspects about the recording, like a dog barking in the background. This will remind me later on if I need to do some extra editing, for example.

Record Your First Episode: Now It's Your Turn

Check off the box beside each task as you complete it.

❏ 1. Introduction planning: 10 minutes.

Write down your episode introduction. Practice it if necessary.

❏ 2. Recording your episode: 45 minutes.

Record away. Adjust your microphone. Don't forget to press the record and stop buttons.

❏ 3. Post-recording steps: 5 minutes.

Rename the recording, back it up, and jot down important notes.

Daily Standup

Did you complete today's tasks?

❏ Yes
❏ No

If no, what do you need to carry over to work on tomorrow?

_____

What did you learn about your business (or yourself) today that will serve you in the future?

_____

Day 12: Intro, Outro, and Theme Song

The Bookends Every Podcast Needs

An Appropriate Theme Song

There is no set rule that podcasts should have any music at all. But adding some music at the beginning and the end will make your podcasts that much more professional. Make sure that the music is no longer than fifteen seconds by itself. So the question is where you can find this music and how much it'll cost.

Surprisingly, you can find some really good music for free. There are some sources listed below. The key is to search for music with a Creative Commons license. Even in Creative Commons–licensed music, some licenses forbid you from using a work for commercial purposes. So keep an eye out for the licensing terms. With that caveat, here are some sources.

- Pixabay: Copyright-free stock music by a community of creators. My favorite source for royalty-free music.
- YouTube Audio Library: This is my second-favorite source. Huge selection of royalty-free music.

- Incompetech: Wide-array of tracks created by solo artist Kevin MacLeod.
- The Free Music Archive: Expansive, free music library for podcasters.

And here are some paid music sources.

- AudioJungle: Browse thousands of free songs or purchase a unique track. Prices start at $1.
- Storyblocks: A $15-per-month subscription for plenty of music and other audiovisual content.
- Shutterstock: A $19-per-month subscription with an even greater selection than the previous two platforms.

> "We launched the show with a twice-weekly format and quickly moved it to weekly. The show has never been a source of income for me, and I found anything more than weekly to be too much to handle. In August 2020, while many podcasts were increasing their podcasts from weekly to daily shows, I decided to move mine to monthly. I worried about our downloads decreasing, but they've actually increased during this time."
>
> —Kathe Kline, Host, *Rock Your Retirement*

Grab Your Audience with a Powerful Intro

What is an intro? Short for *introduction*, your intro is your promise to the listener about your podcast and this particular episode. It's a combination of your music and voiceovers to introduce that episode. These are the elements you should consider for your intro.

1. **Podcast name**: This is a must. Your intro should actually say the name of the podcast.
2. **Tagline**: What is the podcast about? Key value proposition.

3. **Host names**: Say the host names and any qualifications if necessary.
4. **Episode title**: What is the episode about? It could be just the introduction of the guest.
5. **Sponsor names**: If you are sponsored, this is where you introduce them.
6. **Most interesting part of the episode**: This could be the teaser part of the episode to hook the listeners.

Write an intro script and either record it yourself or have it done professionally. I have decided to use one intro for all episodes, but some high-end podcasts create a unique intro for each episode. It's your choice.

Close with a Purpose

The "outro" or closing of the podcast is no different from closing a speech. After thanking the guests and listeners, leave the listeners with a call to action. Examples?

1. Leave a review on the podcasting directory. Unless you ask, they won't give you one.
2. Links to a website or Facebook group, a download location, or a subscribe option.
3. Any promotions, giveaways, or courses they can sign up for afterward.

Just like an intro, you can have this prerecorded for all episodes or do a unique outro for each episode.

Intro, Outro, Theme Song: Now It's Your Turn

Check off the box beside each task as you complete it.

❑ 1. Select your theme song: 20 minutes.

Look through the sites mentioned, and select the music that you like for your show.

❏ 2. Prepare your intro: 20 minutes.

First, write down your intro script. If you want to record it yourself, use an audio mixing tool like Audacity or Jukebox and record it. If you want to get it done professionally, send the music and your script to a professional.

❏ 3. Prepare your outro: 20 minutes.

Write down your outro script. Either record it yourself or send it to a professional.

Daily Standup

Did you complete today's tasks?

❏ Yes
❏ No

If no, what do you need to carry over to work on tomorrow?

What did you learn about your business (or yourself) today that will serve you in the future?

Day 13: Podcast Artwork

Listeners Judge a Podcast by Its Cover—What Does Yours Say?

Artwork Specifications

Just like a book is judged by its cover, your podcast cover art is what your potential listeners will use to select your podcast over others as they browse podcast directories. They'll use your artwork to decide if they should read your description, skim your episode titles, or listen to their first episode, making it a critical part of promoting your podcast.

That's why you need to take your podcast cover art seriously. Before we even go into design aspects, we first need to focus on technical specifications. Podcast directories such as iTunes and Spotify are very strict about these specifications.

- Your podcast art should be square. Make it 3000 pixels by 3000 pixels. This way it will look good everywhere, even when scaled down. 1400 pixels by 1400 pixels is the minimum for Apple Podcasts.
- Make it 72 dpi and use RGB colors.
- Save your art as a JPEG or PNG (but JPEG is better).

Make sure your artwork complies with Apple's requirements before submitting to Apple Podcasts. Review their documentation at Help. Apple.com.

Before submitting to Spotify, browse their delivery specifications to make sure you're compliant. Review their documentation at Podcasters. Spotify.com.

> "I think the biggest challenge was not realizing how much time and effort I would need to create content and edit my podcast episodes. I, like most, thought how hard can it be to record and hit send right?! The deeper I got into it, the more picky I became and the more I added onto my plate. There were plenty of days I thought I should just quit and get my time back.
>
> "But there was always something that kept pulling me back, so I joined groups and forums to learn more tricks of the trade and to refocus on what my goals were for the podcast."
>
> —Tanya Fox, Host, *Fox Talks Business*

> "I started with picking a few Facebook groups where I made a few posts. They weren't a direct promotion as you are usually not allowed to advertise what you do, why you do it, etc. But those posts were an opportunity for people to ask questions and give me a chance to share info about what my goal with the podcast is. Then I launched a dedicated community, which really helped boost listenership and exposure."
>
> —Natalie Luneva, Host, *SaaS Boss*

Creating Your Cover Art

You can create your podcast cover art yourself, or you can pay someone else to do it for you as per your guidelines. Here is a list of these options.

Do-It-Yourself Options

- Canva is great for many designs, and it is also an excellent choice for podcast art.
- Tailor Brands is another great option for logos. After asking some questions, their AI-enabled system comes up with an initial design. You can then customize it.
- Adobe Spark, part of the Adobe suite, is another great option for inexperienced designers, and it provides many templates for you to start from.

Paid Options

- 99Designs is a great resource with talented designers from all over the world. Either you can short-list some designers yourself or open your requirements to the community to pitch their designs to you.
- Fiverr is another great resource. Even though it is not a dedicated site for designers, you'll have a lot more choice across the price and quality spectrum.

Whether you go free or paid, make sure that your cover art is the highest quality within your budget.

Tips for Great Artwork

Here are some tips for some great artwork.

- **Keep it simple**: This tip applies to fonts, words, and colors. Don't mix more than two fonts, more than three colors, and more than seven words. You don't want the listener to spend too much time trying to figure out your cover art.

- **Make the design consistent with your brand**: Your podcast should be an extension of your brand. So the colors, the words, and the font should be an extension of your brand assets.
- **Keep clear of copyright issues**: It's better to be safe than sorry. A word in my first podcast title was contested for copyright infringement. Even though we reached an amicable settlement, it was a headache. So steer clear of copyright issues.
- **Test designs on multiple media**: You'll be using your cover art on multiple social media platforms. So make sure that the cover art comes across clearly in multiple sizes and also in thumbnail format.
- **Don't be afraid to use your photo**: I made a decision to use my picture on all my podcast cover art designs. All my podcasts are an extension of my personal brand, and I wanted to make sure that all my listeners associate my podcasts with me.

Create Your Artwork: Now It's Your Turn

Check off the box beside each task as you complete it.

❏ 1. Design and finish your artwork: 60 minutes.

There are not many steps today—just one. Decide if you want to do it yourself or pay someone to design it for you. Spend the remaining time either designing the art or sending out the requirements.

Daily Standup

Did you complete today's tasks?

❏ Yes
❏ No

If no, what do you need to carry over to work on tomorrow?

What did you learn about your business (or yourself) today that will serve you in the future?

Day 14: Podcast Description and Category

How Can I Find Your Podcast?

Your Podcast Description

Aside from your podcast artwork and your show's name, potential listeners use your podcast description to decide whether they'll listen to your show. If your description is bad—or missing entirely—listeners won't know how to evaluate your show. They'll move on and invest their time with another show whose host took a few minutes to let them know what the podcast is about.

A podcast description is a brief text that describes your show. You can write anything you want and use it as a powerful tool to convince your listeners to listen.

Think of your podcast description like the synopsis on the back of a book or inside the jacket. It's the second thing people read after the title. In most cases, the title isn't enough information for the listener to understand what your show is about, so you'll want to take advantage of this extra space.

How important is your description? In The Podcast Host's 2020 Podcast Discovery Survey, they asked 780 listeners about the importance of several front-facing aspects of a podcast. The podcast description came out on top by a good margin.

Additionally, your podcast description is included in your RSS feed. Anyone who displays your podcast (e.g., iTunes, Spotify, etc.) will publish your description as well.

> "Starting a podcast is easy. There's a lot of truth to the joke that everyone and their mother has a podcast. But keeping it going, especially as a one-man shop, is hard. I have to find the guests, book the guests, interview the guests, edit the interview and, personally, I add in an intro and outro, find the sponsors, record the ads, etc. etc. etc. There's so much to do! It's no surprise that if you make it to 25 episodes, you're in the 90th percentile of podcasts."
> —Jay Shifman, Host, *Choose Your Struggle*

> "The best way to stand out as unique is to, well, BE UNIQUE. We're honest with our opinions. We make it clear how we feel about matters. That gives our podcast a clearly defined message. People who disagree with us won't be drawn to us, but the people who do, will WANT to listen to every episode because it's like talking to a friend. We share our opinions, we make fun of each other, and we create fun segments. These things create a genuine atmosphere that will attract our target audience."
> —Sydney Myers, Host, *Dallas Hoops Cast*

Podcast Category

It's important to optimize how your podcast appears in podcast directories like Apple Podcasts (iTunes) and Spotify. One way of doing that is to pick the right podcast categories for your show. Each directory has a different set of categories and how many you can choose. My advice is to focus on your first category.

Any listeners browsing these categories for new podcasts will do so by interest. So if your technology podcast is listed under politics, you're unlikely to get in front of your target audience and attract new listeners.

Podcast directories like iTunes are also becoming increasingly strict on the way their store appears to users. So if your show is listed in a category that's completely irrelevant to its content, you run the risk of being removed from the entire directory.

Many categories also have subcategories. For example, the fiction category has three subcategories as of this writing. They are comedy fiction, drama, and science fiction. It can boost your show's visibility to choose a subcategory as well as the primary category.

Description Template

The best podcast descriptions are written for humans. They are clear, succinct, and self-explanatory, and they avoid unnecessary repetition.

Even though the podcast directories give up to 4,000 characters for descriptions, you should try to front-load in the first 200 or so characters as most truncate it with "read more" after that.

Additionally, don't forget about search engine optimization (SEO). It can have a big impact on your show's discoverability when people search on search engines like Google. So try to incorporate key terms that listeners may be using to search.

Here is a three-sentence template to get you started.

First sentence: Something your listeners already believe is true and/ or agree with. Your description starts off by acknowledging a problem they have.

Second sentence: What your listeners can expect to hear in each episode. What is your podcast about? What is the frequency of release? What is the format?

Third sentence: Who the show is for, including keywords for search. Mention the types of audiences your show is meant for and any search terms they are familiar with.

Description and Category: Now It's Your Turn

Check off the box beside each task as you complete it.

❏ 1. Describe your podcast: 30 minutes.

Using the template mentioned above, start off by composing your first draft. Given that you are on Day 14 with most of the information already in your hand, this should be a breeze. You know what it is about, who it is meant for, what key problems your show may be addressing, and so on.

❏ 2. Research podcast category: 20 minutes.

Browse through the iTunes podcast categories and select the most appropriate category and subcategory.

❏ 3. Browse through other podcasts in your category: 10 minutes.

Take the remaining time to browse through other podcasts in your chosen category. Is your show in line with those other podcasts?

Daily Standup

Did you complete today's tasks?

❏ Yes
❏ No

If no, what do you need to carry over to work on tomorrow?

_____

What did you learn about your business (or yourself) today that will serve you in the future?

_____

Day 15: Editing Your Podcast Episodes

Get It Ready to Export

Prepare All Your Media Files

Now that you have an episode or episodes recorded, it's time to edit the episode and get it ready for export to podcast directories. The very first step is to get your intro file, outro file (Day 12), any sponsor files, and your raw recorded file. Your specific workflow may vary depending on whether you are editing both video and audio files or just the audio file.

My most recent podcast is also a videocast, so I use Camtasia software to first edit the video file. I load all my intro, outro, and raw files into Camtasia and arrange them in the specific order I want. If I were just editing an audio file, I'd use Audacity software to arrange these files in the order I want. If you've recorded multiple speakers on different tracks, make sure that all the tracks have been loaded. Irrespective of the particular software you've selected, this step should be the same for all. I suggest you have a standard routine to save you time as you move forward.

> "The most challenging part of the podcast is the time it takes to secure and vet guests and then edit the recordings. The actual interviews are always fun!"
>
> —Brooke James, Host, *The Grief Coach*

> "There are two of us that host the episodes, and we are never in the same state since we travel a lot as RVers. Additionally, our guests are not usually near one of us. We have experimented with many platforms and found that Zoom provides the best quality so far. Also, we put in many hours editing, so the sound quality is as good as possible."
>
> —Sean Chickery, Host, *Beyond the Wheel*

Edit for Clarity and Flow

This is where the proverbial rubber meets the road. Here are some basic tips to follow to get optimal quality without spending multiple hours.

1. Don't aim for perfection. Editing is an area that can take a lot of time if you are aiming for perfection. My goal is "good enough."
2. Edit for content first. Make sure to cut any unwanted content and ensure that the flow is logical.
3. Then edit for distractions. Then edit away long silences, speakers talking over each other, and uncomfortable *ums* and *ahs*. My philosophy is to keep the natural *ums* and *ahs* in and not try to be perfect. It's your choice.
4. Make a checklist for audio quality editing. I go through a three-step process of reducing noise, normalizing all the audio, and equalizing my files. You can prepare your own standard checklist.
5. As a last step, insert fade-ins and fade-outs between music and conversation for a smooth professional output.

That's it. I don't try to make it too fancy. Remember, my goal is "good enough." Many people are intimidated by the laborious process of editing. If you are intimidated or don't have time at all, you may want to consider outsourcing the editing.

Add Episode Metadata and Export Final File

As a last step, insert your episode metadata before saving the file. So what is the metadata? Some people call it ID3 tagging. This is nothing but the episode title, track number, podcast name, et cetera. Even though some people may say ID3 data is optional, I think you are better off inserting this metadata so search engines know how to identify your episodes. In Audacity, you'll be presented with a screen to enter this metadata when you hit export.

After entering the metadata, you'll have the option of exporting your output file as a wave file or MP3 file or similar options. You can pick either one. You'll also have the option of providing a bit rate for encoding the file. The recommended bit rates are 96 kbps for spoken word and 192 kbps for music.

Edit and Export: Now It's Your Turn

Check off the box beside each task as you complete it.

❏ 1. Prepare your files: 10 minutes.

Load your intro, outro, sponsor ads, and raw files into your selected software program. To save time, you can add your intro and outro files to your library in the software if they stay the same. Then arrange the files in the order you want.

❏ 2. Edit the episode: 40 minutes.

Start editing your files for content flow and eliminating distractions. Fade in and fade out where necessary.

❑ 3. Insert episode metadata: 10 minutes.

Insert ID3 metadata and export your file as an MP3 file or a wav file.

Daily Standup

Did you complete today's tasks?

❑ Yes
❑ No

If no, what do you need to carry over to work on tomorrow?

What did you learn about your business (or yourself) today that will serve you in the future?

Day 16: Episode Transcripts and Show Notes

How Can I Find Your Podcast?

Episode Title and Description

In addition to your overall podcast description, you should also set aside time to write a brief but engaging episode title and description. The title, description, transcript, and show notes will all help with search engine optimization.

Here are some tips for titles and descriptions.

- Abbreviate episode numbers if you use them (e.g., Ep1 instead of Episode 1) or, better yet, just use numbers. I personally don't number my episodes.
- You have limited real estate for your headings, so get to the point quickly.
- Use long-tail search keywords (e.g., Building a profitable drop-shipping business in thirty days).
- Instead of giving all the secrets in the title, tempt the prospective audience to click to find more. No, I am not talking about "clickbait" headlines but something mysterious.

- Start the description with the most important points of the episode. Many podcast directories will only show the first one or two sentences of the description, so avoid starting with phrases like "In this episode, you'll find . . ."
- Feel free to "drop the names" of your guests in your title and description, especially if they are influencers.

> "We decided to transcribe all our podcasts so they are available to read on our blog. This also helps in gaining new listeners through search results on Google, etc. and we offer downloads for each episode such as checklists, guides, etc., which builds our list, and we notify all subscribers of new episodes via email, too, as well as across all of our social channels."
>
> —Dawn McGruer, Host, *Dawn of a New Era*

> "I started my first podcast in 2014, and I was completely starting from scratch. I attended the NMX Live conference in Jan 2014 and met a ton of amazing podcasters who gave me an action plan in a matter of days to get my podcast launched. Launching my current podcast was fairly effortless because I built off the learning curve of my first podcast."
>
> Stephen Warley, *Life Skills That Matter*

Episode Transcription

A podcast transcript is a word-for-word account of your audio recording. A transcript will help you expand your search traffic, provide backlink opportunities, and create content that you can repurpose on your blogs or elsewhere.

You can create the transcript in many different ways.

1. You can create the transcript yourself by listening to the episode. This is very time consuming.
2. You can use an automated service to transcribe your podcast. Some of the popular services are Trint.com, Temi, Descript, Sonix, and Podscribe. These services vary in quality but are quite inexpensive. For a complete list of providers, join the members-only section at https://rameshdontha.com/members-options/.
3. You can outsource your podcast for human transcription or a combination of human and automated service. Some providers are GoTranscript.com, Rev.com, and Scribie.com.
4. You can also outsource your transcription to freelancers on Fiverr or Upwork, for example.

Depending on your budget and time, you can select one of the above options. My recommendation is to invest in preparing the transcripts as they'll help with show notes and blog articles.

Episode Show Notes

Show notes are an extension of your episode summary, highlighting key points of the episode. Typically, the episode summary shows up on podcast directories along with the title, but show notes are displayed on the website. Show notes also give you an opportunity to include other relevant resources for an in-depth look. Like transcripts, you can also outsource the creation of show notes.

Here is a show notes template for you to use.

Episode title: Remember to make the title SEO friendly and include target keywords in the URL.

Episode summary introduction: Brief paragraph describing the high-level points of the episode. Remember to include the SEO keyword targets relating to the episode's topic in addition to any guest names.

Topics discussed in this episode: Bullet point list of the main episode takeaways. Think of the topics as headlines or questions answered during the episode. Remember to include timestamps when each topic is introduced.

- Topic 1 headline [timestamp]
- Topic 2 headline [timestamp]
- Topic 3 headline [timestamp]

Resources mentioned in this episode: Bullet point list of links to external resources mentioned throughout the episode. Include time-stamps when each article is introduced.

- Link 1 [timestamp]
- Link 2 [timestamp]

Episode transcript: Transcript of the episode's audio.

Calls-to-action: Collect all current calls-to-action together. For example:

- Subscribe to my newsletter.
- Subscribe to Patreon.
- Purchase tickets to next live event.
- Join the podcast's Facebook group.

Description, Transcript, Show Notes: Now It's Your Turn

Check off the box beside each task as you complete it.

❏ 1. Describe your episode: 10 minutes.

Use the podcast script you've already prepared to give your episode a title and a description. Remember to use search engine keywords.

❏ 2. Prepare transcript: 30 minutes.

If you've outsourced your transcription, you don't need thirty minutes. If you use an automated transcription service you may want to give the transcript a quick review to fix major issues.

❏ 3. Prepare show notes: 20 minutes.

Use the transcript to highlight key points from the episode. For speaker bios and resources, I suggest that you get this information when you are vetting the speakers. Put them all together using the template provided above.

Daily Standup

Did you complete today's tasks?

❏ Yes
❏ No

If no, what do you need to carry over to work on tomorrow?

What did you learn about your business (or yourself) today that will serve you in the future?

Day 17: Hosting Your Podcast

Starting to Distribute Your Podcast

Selecting Your Podcast Host

You can't upload your podcast file to Apple Podcasts or Spotify. They are only listening platforms. Instead, you should upload your files to a podcast host such as Buzzsprout, Blubrry, SoundCloud, Captivate, Transistor, Castos, et cetera. The listening platforms such as iTunes get the audio file via an RSS feed from the podcast host.

So the question now is which podcast host. As podcasting has grown at a breakneck speed, so has the number of podcast hosts. There are hundreds of hosts to choose from, and more are getting into the space.

I'd look for four main things in a host.

1. Storage capabilities. Over a period of time, you need more storage to host your increasing number of files. So price is one factor.
2. Audience analytics. After you launch your podcast, you want to know who is listening and from where and more sophisticated analytics. Some hosts charge for advanced analytics.
3. Website integration. More than likely, you'll have a website with your blog along with links to your podcast files. Some

podcast hosts have an easy-to-use integration with your website, which makes it easy to upload your files.

4. Marketing and promotion. Some podcast hosts provide other features such as transcription service, show notes service, and audiograms for promotion. These additional services can be handy when promoting.

My top five choices are Castos, Blubrry, Buzzsprout, Captivate, and Transistor. You can check out the https://rameshdontha.com/members-options/ member section for a much more comprehensive listing.

> "My next challenge was knowing which platform to host on. I took to Google and read reviews. I chose Podbean, but in hindsight I wish I'd chosen Anchor because the landing pages for each episode on Anchor link to several different podcast apps (Apple, Google Podcasts, Spotify), thus making it easier to share your podcast to audiences who use different apps to listen."
>
> —Helen Croydon, Host, *The Media Insider*

> "The biggest challenges were updating the RSS feed from my previous podcast, as well as updating the icon (artwork of the podcast). Meeting the criteria for each was quite difficult. My podcast partner and I are still new at working together, so we are finding our rhythm. We seem to have gotten our legs and now have people inquiring about advertising."
>
> —Jeremy Nunes, Host, *Dynamite Drop-In*

Signing Up with a Podcast Host

Once you've selected your podcast host, signing up with them is pretty straightforward. You need to pick a service that best fits your budget and needs and provide the necessary information about your podcast.

The material you've prepared so far regarding podcast name, description, and artwork are all needed to set up your account with the podcast host. So please take additional time to make sure that all of them are current and accurate.

Uploading Your First Episode

The established podcast hosts have made the onboarding relatively easy. You'll have a couple of options to upload your episodes to the host. One option is to directly upload via their website. The other option is upload-ing via your website with a plug-in provided by the host.

The episode title, the description, and the audio file are the bare minimum of what you need to upload your episode. It is always rec-ommended that you start your podcast with multiple episodes instead of just one.

In the subsequent sections, you'll learn how to distribute these epi-sodes to directories such as iTunes and Spotify.

Selecting Your Host: Now It's Your Turn

Check off the box beside each task as you complete it.

❏ 1. Select your podcast host: 20 minutes.

You'll not go wrong with any of the top-rated podcast hosts. Select the one that fits your budget and your requirements.

❏ 2. Set up your account: 20 minutes.

Set up your account with your selected host.

❏ 3. Upload your episodes: 20 minutes.

Upload your initial launch episodes.

Daily Standup

Did you complete today's tasks?

❏ Yes
❏ No

If no, what do you need to carry over to work on tomorrow?

_____

What did you learn about your business (or yourself) today that will serve you in the future?

_____

Day 18: Website for Your Podcast

A Quick and (Not Really) Dirty Website

Selecting the Domain Name and Website Host

I am not a big fan of a fancy website for a podcast during the initial days. My recommendation is to have a website as a reference and to keep your show notes and transcripts along with links to your podcast files on directories like iTunes. And that is the reason I am not allocating any more than sixty minutes for this task. Sure, you can take multiple days or even weeks to get a fancy website ready. It's your call.

The very first thing you need is a domain name and web host (just like your podcast host). You can purchase domain names from merchants like GoDaddy and Namecheap. And there are plenty of web hosts who actually give you free service for multiple months. Or you can get a free web hosting from WordPress, for example. I suggest you spend as little time as possible on domain name purchasing and web hosting.

"Our 3rd month of the podcast was when we really started to see our numbers grow, and we started to reach across to more than 20 countries. This was really exciting as it was great to see our reach grow but also because we still to this day receive emails from people around the world saying they listen in. It is a wonder that in today's world I have had the honour to interview people from Australia, Singapore, UK, Poland, and Africa. I cannot imagine a time years ago when this would be something that would have happened. I think going global is something that every business strives for. Podcasting, I feel, gives you that avenue much quicker as well as helps you to create amazing relationships around the world that are easier than ever to keep cultivating."

—Tanya Fox, Host, *Fox Talks Business*

"I sought out the help of a mentor that had forged a successful podcast for himself and his business. It was with his guidance that I paused to make sure I knew who I was trying to reach, what my show was going to be about, where I was going to take them as well as why they should listen to me vs. all the other podcasts out there. I believe it was this clarity help me launch as successfully as I did, and quickly be ranked in the top 100 charts in four countries."

—Tracy Brinkmann, Host, *The Dark Horse Entrepreneur*

Selecting a Website Template

The next step to get your website ready is to pick a content management system (CMS) and get a theme installed. WordPress is the most popular CMS, and approximately one-third of all websites run on WordPress. Of course, you can go with other CMS such as Wix or Joomla as well.

Once you've narrowed down on a CMS, shop for a template that is optimized for podcasts. There are many free and paid templates that are specifically made for podcasts. For example, a template called "podcast" is free for WordPress websites. So you get both the web hosting and the template for free if you go this route. You can literally spend less than thirty minutes to get your website up and running as many web hosts provide a five-minute WordPress installation along with a template.

Populating the Website Content

This is the step you may want to spend some time in getting your website ready. The minimal content you want to have ready is the podcast description, intended audience, contact information, and "About" section for the host(s). If you have been following the day-to-day steps, you should have all that information ready by now. All you have to do is populate the website with the information you have been preparing all along.

Additionally, provide information on launch episodes, links to podcast directories (if you have them at this time). We'll discuss podcast directories in the next chapters.

Setting Up Your Website: Now It's Your Turn

Check off the box beside each task as you complete it.

❏ 1. Select your domain name and web host: 15 minutes.

Go to a merchant like GoDaddy or Namecheap, check for available domain names, and buy the most relevant domain name for your podcast. Then sign up with a website host with an appropriate service.

❏ 2. Select the website template: 20 minutes.

Search for a website template developed for podcasters. If you have decided to go with the WordPress CMS, there should be plenty of free templates readily available.

❏ 3. Prepare the content: 25 minutes.

Log in to the back end of the website (administration area) and update the information about your podcast, contact info, "About section," and initial episodes.

Daily Standup

Did you complete today's tasks?

❏ Yes
❏ No

If no, what do you need to carry over to work on tomorrow?

_____

What did you learn about your business (or yourself) today that will serve you in the future?

_____

Day 19: Getting Listed in Podcast Directories

Directories Are Key to Wide Distribution

Introduction to Podcast Directories

I talked about podcast hosts (Buzzsprout, Blubrry, Castos, etc.) in an earlier chapter. They host your audio files. But the listeners get their episodes from podcast directories like Apple Podcasts (iTunes), Spotify, Stitcher, and Google Podcasts. Apple Podcasts is the largest directory of all, with more podcasts listed and more listeners accessing the podcasts from them than from any other directory.

So what do you do? Step 1: Select one podcast host (Buzzsprout, Blubrry, Castos, etc.) and upload your files to your selected host. Step 2: Register with as many podcast directories as you can, and submit the RSS feed from your podcast host to these directories.

Then you'll see the magic happen. Every time you upload a new episode to your host, the directories will automatically get those episodes from your host and start distributing. Once you register and submit your RSS feed initially, there is nothing for you to do afterward.

"You have to have a launch plan at least six weeks in advance to prepare branding, which is so important, and to ensure you are using a host that allows growth and tracking as well as easy distribution across the obvious podcast players such as Apple, iTunes, Google Podcasts, Spotify, etc."

—Dawn McGruer, Host, *Dawn of a New Era*

"The first challenge was setting up the infrastructure of the podcast (podcast hosting platform and a website that was easy to update with podcast episodes and which accepted a feed from the podcast hosting platform). I tackled this by hiring a company that sets up podcast frameworks."

Bryan Smith, Host, *DreamPath*

Focus on Top Podcast Directories

As of the initial publication of this book, there are thirty-plus podcast directories. So which one should you focus on? My recommendation is to submit to as many directories as you can, starting with Apple, Google, Spotify, and Stitcher. These four directories have the most market share.

Below are key requirements for these top four directories. Please check the members-only section on https://rameshdontha.com/members-options/ for detailed instructions.

Apple Podcasts: You need an Apple ID and to go through the iTunes Connect site to register your podcast. The most important thing you need is the RSS feed link from your podcast host. Apart from this, you need the title, description, category, artwork, and website as well. You may have to wait up to a week to get your podcast approved, so plan accordingly.

Google Podcasts: Google has a fairly straightforward process. The most important item you need is the RSS feed link.

Spotify: Spotify started with music streaming, and they have become a major player in podcast distribution right behind Apple. To submit your podcast to Spotify, you need the title, description, RSS feed, and at least one episode. This episode can be a trailer and not a fully recorded episode.

Stitcher: Stitcher is the smallest player of these four but a growing distributor. Stitcher is only a mobile app, and it can be found on mobile devices only. The process is very similar to the other directories, so it should be very straightforward.

Getting Listed on Tier 2 Directories

So who is next? Quite a few but it's up to you. If you have time, please go ahead and submit to as many as you can. Here are some of the other prominent directories.

- iHeartRadio
- Castbox
- Pandora
- Overcast
- TuneIn Radio

Apart from these, your podcast host may also have a podcast directory like Blubrry. So go ahead and get it listed there as well.

Getting Listed in Directories: Now It's Your Turn

Check off the box beside each task as you complete it.

❏ 1. Submit to Apple, Spotify, Google, and Stitcher podcasts: 40 minutes.

Prepare the required information, such as title, description, categories, artwork, et cetera. Get your RSS feed from your podcast host. Submit to at least these top four directories.

❏ 2. Submit to other directories: 20 minutes.

Submit to other directories if time permits.

Daily Standup

Did you complete today's tasks?

❏ Yes
❏ No

If no, what do you need to carry over to work on tomorrow?

_____

What did you learn about your business (or yourself) today that will serve you in the future?

_____

Day 20: Launch Your Podcast

Time to Open the Curtains

Types of Launch

Just like a movie or a book launch, you can have a grand launch or a soft launch. Let's talk about both types of launches and the pros and cons of each.

A grand launch: A grand launch typically involves picking a date, reaching out to friends and family, reaching out to other podcast hosts and influencers, and possibly a press release if you are in the mood. I picked this type of grand launch for my second podcast and focused on LinkedIn as the preferred social media platform to launch. I kept teasing with a trailer, upcoming guests, and key episode nuggets.

A soft launch: If you are not confident about the launch and want to build momentum gradually, you can go this route. You could just start publishing to directories and possibly reach out to a few very close friends or followers. This is a less stressful way to get your podcast out there and possibly correct any mistakes in the first week or two.

In my opinion, there is no right or wrong method. Pick the one that best suits your needs.

> "The biggest challenge for me was the mind drama of my inner critic with questions like 'What if no one likes the podcast?'; 'What if they make fun of me?' So I needed to remind myself that there is someone out there that really needs to hear this message, and my discomfort is worth it to get the message out to that person. Shifting away from being worried about me and instead loving on my people and being in service to them."
>
> —Danielle Savory, Host, *It's My Pleasure*

> "*Sweet but Fearless* podcast has three cofounders, and we use Zencastr for free to record a lot of our episodes and then find a Fiverr employee to edit and post for us on all mediums. We then bought a Zoom H6 recorder and mic to allow for better sound and editing. The main hurdle is always the belief that you sound OK, will people like your voice versus your message. We now have regular guests on our podcast and showcase women and their career transition journeys."
>
> —Mary Sullivan, Host, *Sweet but Fearless*

Activities for Launch Day or Launch Week

The following are some of the activities you can do for your launch.

Prepare a teaser episode. This can be a trailer or excerpts of interviews from your episodes. This will also help you gauge interest in your show.

Record and upload multiple episodes. This is some of the best advice I got before I launched my podcast. The ideal number for launch day is three episodes. Your listeners will know that you are serious about your podcast and will be more likely to subscribe to your future episodes.

Announce on your favorite social media platform. My favorite social media platform is LinkedIn. I made multiple posts during my launch week, profiling the guests and releasing the trailer. Whatever your favorite media is, post multiple times.

Reach out to other podcast hosts. This is something you could be preparing for before the launch week. Reach out to other hosts and ask them to announce your launch. Generally, podcast hosts want to help each other out.

Get the Message Out

Now that you've gotten the launch butterflies out of the way, start spreading the message. We'll talk more about promotions in subsequent chapters, but it's never too early to promote your podcast.

If you have an email list, send out an email blast announcing your show.

If you are a member of any Facebook groups, post there. Please do follow FB group guidelines as some admins might reject self-promoting posts. Try to add value.

Use other communities on Reddit and Quora. There are some great podcast communities on Reddit, and they'll be eager to provide you feedback.

Guest post on other blogs. There are some blogs that allow guest posting with one or two links to your podcast.

Launch Your Podcast: Now It's Your Turn

Check off the box beside each task as you complete it.

❑ 1. Check all engines are ready to go: 20 minutes.

This is the time to make sure that your episodes are uploaded to your podcast host, your website is ready, and the podcast directories are showing your show. Take time to check these important items.

❑ 2. Decide on the type of launch: 20 minutes.

Decide on a grand launch or a soft launch and act accordingly.

❑ 3. Send the message out: 20 minutes.

Start sending out messages, posting, and blogging away. It's showtime!

Daily Standup

Did you complete today's tasks?

❑ Yes
❑ No

If no, what do you need to carry over to work on tomorrow?

_____

What did you learn about your business (or yourself) today that will serve you in the future?

_____

Day 21: Promotion—Friends, Family, and Network

Start with Your Inner Circle

Defining Your Inner Circle

First of all, pat yourself on the back. Getting this far is not easy. Twenty days ago, you didn't have a podcast. Now you do. And it's not an easy thing to do. But you did it!

If you are proud of yourself, most likely there are other people who are proud of you as well. These are your most ardent and unconditional supporters who have been encouraging you all along. Reach out to them. Thank them for their support and ask them to listen, subscribe, and download your initial episodes. Podcasting is a relatively new medium, and supposedly, the average podcast episode gets a hundred listeners. That's it, just a hundred. So every listener counts.

Also remember that every listener will more than likely share the information with at least one other listener. So get your inner circle of friends, family, and colleagues to listen to your initial episodes. Don't forget to ask them for feedback as your interest is in improving the quality of your podcast with every episode.

"Eventually, as the podcast audience grew and the show established itself as a source of quality content for our subject, more and more guests began pitching themselves, or were pitched to us through major publicists. So now we're in a kind of self-reinforcing positive feedback loop where quality guests increasingly want to be on the podcast, and that in turn is driving the show's promotional exposure. The key takeaway is if you focus on quality and integrity, the audience will start showing up on its own thanks to word of mouth."

—Guy Nadivi, Host, *Intelligent Automation*

"When we first started *FIT CHICKS Chat* we were uncertain if anyone would ever listen. Like anyone who has fears about putting themselves out there, we were not sure what would happen but we went for it as we knew this was our favorite way to share our message vs social media. To date we have been rated one of the top health and fitness podcasts on iTunes (top 100 list) and have over 23k downloads every 90 days."

Amanda Quinn, Host, *FIT CHICKS Chat*

Be a Guest and Network

Podcast hosts are an interesting community. I noticed that they help and support each other. I have spent numerous hours preparing a list of interesting podcasts from iTunes and Spotify and reached out to the hosts about my podcast. I asked them if they'd be interested in being a guest on my future shows (if there is a match) and if I could be a guest on their show.

There are some good sites where you can find a list of podcasts. Here are a few.

- Radio Guest List
- Reddit group for podcast guests

- Facebook group—Podcast movement
- Facebook group—Podcast support group

Don't Forget Reviews

Podcast reviews and ratings can be very helpful in the long run. Apple Podcasts' "New and Noteworthy," which can be a great free promotion, will be looking for reviews in addition to downloads.

Here are some tips to get reviews:

Ask your guests to review their own episode. They'll more than likely oblige.

Ask your listeners (your inner circle, for sure) to review the podcast on iTunes, Spotify, et cetera.

Unfortunately, posting a review on iTunes is not a straightforward process. So I wrote a post about how to review on iTunes and shared it with my listeners. You may want to do that as well.

In general, don't be bashful about promoting your podcast. If you don't talk about it, nobody else will.

Friends, Family, and Network: Now It's Your Turn

Check off the box beside each task as you complete it.

❑ 1. Put together a list of your inner circle: 20 minutes.

Prepare a list of your family and friends who will support your podcast. Also, put together a list of podcast hosts that you plan to reach out to.

❑ 2. Send emails and reach out: 20 minutes.

Start sending out emails or start calling people. Put focus on Apple Podcast downloads initially as that is the main source for many listeners. Give them the link where they can download the episodes.

❏ 3. Reach out to other podcast hosts: 20 minutes.

Reach out to other podcast hosts, either with an invitation to be a guest or an offer to be a guest yourself.

Daily Standup

Did you complete today's tasks?

❏ Yes
❏ No

If no, what do you need to carry over to work on tomorrow?

_____

What did you learn about your business (or yourself) today that will serve you in the future?

_____

Day 22: Social Media: LinkedIn, Twitter, Instagram

Prioritize Your Social Media Platforms

LinkedIn Strategies

There are so many social media platforms out there. And there is only so much you can do. In this chapter and the next few chapters, I'll be covering promotional tactics on as many social platforms as possible. It does not mean that you have to work with each one of them. Given that social media is such a crucial medium, each of us should focus on at least one platform. My preferred go-to platform is LinkedIn. Please feel free to connect with me there.

Here are some tips to promote your podcast on LinkedIn. Please bear in mind that this section is not about growing your LinkedIn followers but working with existing connections and increasing engagement with them.

- Add the podcast info (and link) in your headline and profile sections.
- Announce that you've become a podcast host as a job change. This generates great engagement.

- Write a LinkedIn post about your podcast and tag some of the guests you've interviewed. If possible, make a "headliner" video with a snippet from your episodes.
- Add relevant hashtags to your post to expand the reach of your post.
- Add one of the posts about your podcast as a featured post, which shows up in your profile.
- Ask your friends and connections (either via email or LinkedIn messages) to engage on your podcast posts so their followers/connections get to see the podcast as well.
- Join some relevant LinkedIn groups and post about some of the things you've learned or some related and useful content. Just don't spam about your podcast.

"As far as promoting, a big portion of it was done via Instagram and encouraging people in our network to like and share. The social media strategy is primarily around Instagram as that is where the highest engagement is. This is a pivot from initially planning to be present across Facebook, Twitter, and LinkedIn as well—but I pivoted to focus on where the engagement is. At the end of each episode, we encourage people to rate and review on Apple Podcasts as that helps other people looking for this type of content find it."
—Brooke James, Host, *The Grief Coach*

"I promoted my podcast primarily on my social media. Particularly responding to comments and asking fan questions through Instagram stories for content was key, as people are more willing to listen if they feel included. The biggest thing I've found is staying consistent: I've built credibility through my podcasts by releasing every week and slowly building a following base."
Polina Edmunds, Host, *Bleav in Figure Skating*

Twitter Strategies

You might find some of the tips common for all social media because they work on all platforms. Here are some Twitter-specific strategies I have followed, and I continue to do so to expand the reach of my podcasts.

- Add your podcast name and link to your Twitter profile.
- Add your podcast trailer or an interesting episode as a pinned tweet.
- Share multiple tweets for a single episode with different captions.
- Tweet with an audio file or video "headliner" file for each episode.
- Use relevant hashtags with your tweets.
- Find relevant guests on Twitter by searching for topics and hashtags.
- Don't forget to tag/mention your guests in your tweets.
- Tweet with behind-the-scenes activity about your podcast recording.
- Tweet about your past episodes as well.

Instagram Strategies

Instagram is a totally different platform from LinkedIn and Twitter. It is a visual platform, so you need to plan accordingly. I'll be honest: I don't focus much on Instagram. So my Instagram tips are based not on my experience but on the feedback from Instagram-influencer guests I have interviewed for my podcasts.

- Jazz up your bio. That's the first thing Instagrammers see when they come to your feed.
- Link your podcast website to your bio.
- Take selfies with your guests and post by tagging them. My cohost is really good at this.
- Create audio story highlights and posts. Once you have ten thousand followers, you can add a link in your story.
- Post audio snippets.

- Post quotes from your guests.
- As Instagram is a visual platform, make sure your feed is visually appealing.
- Follow other podcasters and influencers.

LinkedIn, Twitter, Instagram: Now It's Your Turn

Check off the box beside each task as you complete it.

❏ 1. Select your preferred social media platforms: 10 minutes.

Hopefully, you've an idea about your preference based on the followers and engagement you get.

❏ 2. Create audiograms and videograms: 30 minutes.

There are many tools like Headliner.app to make audio or video snippets. Also, use Canva to create images with guest quotes or selfies.

❏ 3. Post and engage: 20 minutes.

Update your profiles, post the snippets, and engage.

Daily Standup

Did you complete today's tasks?

❏ Yes
❏ No

If no, what do you need to carry over to work on tomorrow?

What did you learn about your business (or yourself) today that will serve you in the future?

Day 23: Social Media: Facebook and Facebook Groups

Vast Reach with Extensive Engagement

Facebook Pages

Facebook is ideal for promoting podcasts for a couple of reasons. (1) Facebook is great for sharing different kinds of content like videos, audios, text, gifs, and images. So you can experiment with different types of promotional content. (2) It's the largest social media platform out there with an average thirty-three minutes spent on there per day by each FB user.[10] In other words, every one of your potential listeners is out there.

So the question is, how do you reach them? One of the first things you need to do is create a Facebook page because you can't promote your personal page. Set up a Facebook page dedicated to your podcast with great imagery and extensive details in the "About" section.

---

10   H. Tankovska. "Average daily time spent on selected social networks by adults in the United States from 2017 to 2022, by platform," Statista, April 28, 2021, www. statista.com/statistics/324267/us-adults-daily-facebook-minutes.

Facebook rewards people who post consistently and who drive engagement with likes, shares, and comments. What is in your immediate control is posting consistently. A content calendar can be very handy to help you post consistently. With respect to the interesting content that drives engagement, it may take some time for you to come up with that. But don't post 100 percent self-promotional content as it has not proven effective in driving engagement. Post interesting quotes from your guests, informational resources, and inspirational material to start with.

There are some content schedules like Buffer, Hootsuite, Sprout Social, and CoSchedule that can help you with scheduling your content for your podcast page.

> "We stream the podcast live to YouTube, Facebook, and Twitch. Then we distribute the audio file to necessary platforms and additionally convert the podcast into text form for SEO. Micro videos, quotes, and posts are made and then redistributed to Facebook, Instagram, and LinkedIn."
>
> —Jack Fleming, Host, *Process over Profit*

> "I actually don't promote my podcast directly. Any paid promotion I have ever done is actually through gaining notice on social media for myself first. Number two, I am naturally active on social media making acquaintances, replying to people, using it for work in the DMs, talking about more than your standard self-promotion career posts, and people learning more about me. From there, people who want to join in on listening to my film review podcast can and do."
>
> —Nicole Russin-McFarland, Host, *CinematNIC*

Facebook Groups

I discovered Facebook groups while writing my first book, *The 60 Minute Startup*, and I have been in love with them ever since. Since then, I have also started a couple of FB groups myself. There are tens and hundreds of FB groups for every topic under the sun.

Every FB group has its own rules about posting promotional content. First, read and follow those rules. The best way for people to get to know you is by sharing useful and educational content. For example, you can post about the trials and tribulations of starting your podcast in relevant groups and add what you've learned. The more you engage on your page and in these groups, the more chances there are for your audience to get to know you.

Content and Engagement

As mentioned above, one of the advantages of FB is that you can post different kinds of content. Still, videos and images rule the roost when it comes to FB algorithms promoting your content. Lately, FB algorithms prefer video content. So take advantage of that and drive more of the content that FB prefers.

One of the other ways to drive engagement for your content is to collaborate with other podcasters and influencers. Either you can search for other podcast hosts/influencers in your category or set up Google Alerts to find your influencers. Once you've identified some, send out messages to cross-collaborate content.

Facebook Promotion: Now It's Your Turn

Check off the box beside each task as you complete it.

❑ 1. Set up Facebook page: 20 minutes.

Setting up a FB page is a pretty straightforward process from your personal page. Make sure that you've got great imagery.

❑ 2. Research Facebook groups: 20 minutes.

Facebook groups can be a gold mine to find potential listeners and other collaborators. Spend some time researching and short-listing candidates.

❏ 3. Start Facebook engagement: 20 minutes.

Begin the process of posting, engaging, and contacting other people. This is just the beginning of the process. I am not expecting you to complete all of this today.

Daily Standup

Did you complete today's tasks?

❏ Yes
❏ No

If no, what do you need to carry over to work on tomorrow?

What did you learn about your business (or yourself) today that will serve you in the future?

Day 24: Promotion Using Reddit, Quora, Guest Podcasting

More Mediums, More Listeners, More Downloads

Reddit

R eddit is one of the twenty-five most trafficked websites but still not very well understood. I have become a Reddit user myself much more recently. Reddit is a community of communities. You can find your own sub-niche community for any topic on the planet. And that serves well for podcasters. Reddit can be a gold mine for finding topics for your podcast as well. Here are some suggestions for promoting your podcast on Reddit.

- Start with www.reddit.com/r/podcast/. With more than 50,000 users, this community can help you navigate the podcasting world on Reddit.
- Find your own community or subcommunity. Your podcast is about politics? Here you go: www.reddit.com/r/politics.

And you can go sub-niche as much as you want within this community.

- Follow the community guidelines and post content. Make it valuable instead of self-promoting your podcast. Ask questions; offer advice. As you engage more, you'll get your followers.
- Attend and host Reddit Ask Me Anything (Reddit AMA) sessions about your topics.
- Finally, you can also advertise on Reddit if you are up for paid promotions.

"Most of my tips are simple. Work hard. I've worked my ass off finding other podcasts to go on, finding places for both free and paid advertising. Upgrading my mic twice now, learning new editing tools, etc. But you can't do it alone. Places like Facebook groups, LinkedIn, and Reddit have been huge in meeting other podcasters and learning from them. But at the end of the day, content is key. If you're not putting out good content, no one will listen."

—Jay Shifman, Host, *Choose Your Struggle*

"To promote your podcast, get some social influencers to come on your show as their name alone will attract listeners. I have been on radio shows, podcasts, and even national news shows, and almost all of them I push on my social media giving credit to the host. Also work together with other podcasters to interview each other so your listenership will grow. Help one another, especially if you have different shows but some common subject threads."

Frederick Penney, Host, *Radio Law Talk*

Quora

Quora is a question-and-answer community. The great thing about Quora is that the question-and-answer threads are indexed by Google and can be a great source of traffic for your website as well. Here are some tips for Quora promotions.

- Quora is all about thought leadership. So your first goal should be to establish thought leadership by answering questions related to your topic.
- Also experiment with asking questions and use thoughtful questions to engage readers.
- Set up alerts related to your topic so you get notified about questions, and you can answer them right away.
- Quora also allows you to distribute your content. So use this to post informational content with the right keywords so your listeners can find you and your podcast.
- Finally, you can also advertise on Quora. Use appropriate targeted advertising to find your listeners.

Guest Podcasting

You may have heard of guest blogging to get backlinks for your website. Guest podcasting is somewhat similar and can be very effective. Here are a few things you can do to reach out to other podcast hosts to be a guest.

- First, find other similar podcasts starting with iTunes's podcast list (https://podcasts.apple.com/us/genre/podcasts-business/id1321?mt=2).
- Here are some services that can help book you as a guest: PodcastGuests.com, MatchMaker.fm, Podchaser Connect.

- Be proactive and help the host by sending them a concise summary of what you'll talk about, why, testimonials, and any giveaways.
- Send an audio or video request so the hosts know upfront how you'll sound.
- Make an offer to have them as a guest on your show if it makes sense.

Reddit, Quora, Guest Podcasting: Now It's Your Turn

Check off the box beside each task as you complete it.

❑ 1. Reddit: 20 minutes.

Even though I am allocating twenty minutes for each of the three media mentioned here, you may decide to focus on only one of them. So use your time as you see fit. Research the subreddit forums and start engaging.

❑ 2. Quora: 20 minutes.

As mentioned above, start either by posting some relevant content, answering questions in your domain, or ask a question to start engagement.

❑ 3. Guest podcasts: 20 minutes.

Research the services mentioned above and start asking to be a guest on their show. Best of luck!

Daily Standup

Did you complete today's tasks?

❏ Yes
❏ No

If no, what do you need to carry over to work on tomorrow?

_____

What did you learn about your business (or yourself) today that will serve you in the future?

_____

Day 25: Paid Promotions

You Get What You Pay For

Is It Worth Paying for Promoting a Podcast?

Paid promotion is a great way to get your podcast in front of a lot of people who may not have heard about your podcast otherwise. With the right ad at the right place and with the right exit link, you'll be able to tap into your potential new audience and see some massive gains in your listenership. And the easier you make it for people to find your show from an ad, as well as the clearer and more legible you make that ad, the better it'll perform.

In general, paying for advertising your podcast has the following benefits.

- Fastest way to get your podcast in front of your targeted listener.
- Ad targeting will help you widen your audience beyond the core audience.
- Advertising your podcast can differentiate your show from your competition.
- Pay-per-click (PPC) can be a cost-effective campaign.

Of course, it all depends on your budget.

"I created social handles and am investing in Facebook and Google ads, but since this is a passion project and I haven't monetized it, I can't invest more than a nominal amount (a few dollars a day). I try to make up for it with earned media. With each episode, I ask, Who might be interested in this? and pitch it to those folks; for instance, since I released the episode with Dr. Jones on MLK weekend, I sent it to reporters who have covered MLK Day in the past, numerous high school history department chairs, admin at the University of San Francisco (where Dr. Jones teaches), and alumni relations staff at his alma mater (Boston University Law School, Columbia University).

—Winston Chang, Host, *Heard About*

"We've continued with search ads, engaging with the fandom heavily on Twitter and Instagram, pitching directly to academic outlets, and studying what our 'competition' is doing (which wasn't much). Essentially, the success of our podcast was built on so many other promotional activities we invested in that had little to do with the podcast programming itself. We're super excited to continue learning and engaging with even more promotional activities in the future."

Sarah Royal, Host, *Enough Wicker: Intellectualizing The Golden Girls*

Social Media Paid Advertising

As you've read by now, there are myriad social media platforms out there: Facebook, Instagram, LinkedIn, Twitter, Reddit, and Quora, just to name a few. And we didn't even mention Google AdWords, which is another great resource for paid promotions. So which one is it? How to go about it? How much budget? Honestly, I wouldn't be able to address all those questions in detail, but here are some suggestions.

- Choose your platform based on your podcast category and your audience. For example, if your target is business professionals, LinkedIn is better than Instagram.
- Start slow, get feedback, and adjust your campaigns.
- My experience is that Google AdWords and Facebook Ads have been the most effective because I could micro-target my listeners and adjust my ad copy as I learn more.
- Keep focusing on the potential listener "problems and solutions" in your ads.
- Send the clicks to your podcast homepage and not each initial episode page initially.
- Consider a contest to gain subscribers. This works very well with FB Ads.

Advertising on Podcast Apps

Another great way to advertise is with other podcasts, podcast listening apps, and podcast ad networks. Let us discuss each one of them first.

Podcast listening apps are platforms where the listeners download their podcasts. These are Spotify, Overcast, Pocket Casts, Podcast Addict, and Podbay. The advantage with these is that only people who are interested in podcasts would be targeted with these ads.

Podcast ad networks are companies that aggregate ads for various other media such as other podcasts or even other media like print, et cetera. These are companies like Midroll Media, PodGrid, PodcastOne, AdvertiseCast, and Archer Ave. Instead of researching for other great podcasts, these organizations scout for advertising platforms beyond social media.

Lastly, you can reach out to other super-successful podcasts to advertise on them. For example, National Public Radio (NPR) has some very successful podcasts, and you can advertise on them. As with the podcast listening apps mentioned above, the advantage is that you'll be targeting listeners who are already interested in podcasts.

Paid Advertisements: Now It's Your Turn

Check off the box beside each task as you complete it.

❏ 1. Decide on your target platform: 20 minutes.

Based on your budget, time, and existing reach, decide on the type of platform you want to advertise on. Do you want to use social media? If so, which platform? Do you want to use PPC?

❏ 2. Prepare ad copy: 20 minutes.

Start working on your initial ad copy.

❏ 3. Implement your plan: 20 minutes.

Time for action.

Daily Standup

Did you complete today's tasks?

❏ Yes
❏ No

If no, what do you need to carry over to work on tomorrow?

What did you learn about your business (or yourself) today that will serve you in the future?

Day 26: How to Monetize Your Podcast

Why Did You Start Your Podcast?

C ongratulations on reaching this far. Many people give up wayyyy before this step. For the next five days, I'll be writing about different ways of making money with your podcast. Let's focus on making some money!

Go back to Day 1 and check what you wrote in your daily tasks about why you started your podcast. Is your podcast an extension of the products or services you already sell? Did you start your podcast for some passive income stream? Was monetization a goal at all?

Whatever may have been your initial goal, revisit your goals. Now that you have come this far, it is quite possible that your goals have changed. It is OK. Adjust them based on the current situation.

> "As for monetizing the podcast, in the beginning, we were only focused on creating content and building an episode library to show consistency. We were unaware of monetizing immediately, which is something we would encourage others to do. Listenership and subscribers are important metrics to help a podcast monetize; they allow you to leverage ad buys or make money from applications

such as Anchor. If you do not take that route, another way is to pitch to businesses that align with your audience and come to an agreement on what would work best for both parties."

—Ayanna Dutton/Delaila Catalino, Hosts, *Non-Corporate Girls*

"I've just started to monetize my podcast after five years. It wasn't my original goal, but it turned out I'm good at it and like producing content to help people. The focus shouldn't be about making money because it will come across that way. Instead, keep it fun and engaging because it's a considerable investment of time and resources. Also, ensure you have the right equipment to make the process easier. Otherwise, you'll struggle to enjoy the process of making monetizable content for your listeners."

—Nathan Webster, Host, *Let's Talk Marketing*

Common Ways to Monetize a Podcast

Here are some common ways to monetize your podcast. We'll go into each one of these strategies in more detail in the subsequent chapters.

- Another channel to sell your products or services: You can use your podcast to pitch your own products or services. This strategy provides an extra channel to your existing business or new business.
- Affiliate or commission income: You can sell your guests' products or services on your show and get a commission from them. Or another way is to be an affiliate for some other product or service and include their links in your website content.
- Sponsorships and advertisements: This is one of the more typical ways to monetize podcasts by including advertisements at the beginning, middle, or end of your podcasts.

- Subscriptions and listener donations: Another way to monetize your podcasts is to offer a subscription service to your podcasts or ask for listener donations via your website or a service like Patreon.

Not-So-Common Ways to Monetize

- Sell your podcasts or services to your guests: This is an interesting way to monetize, in which you invite your potential clients as your guests and, in the process, become familiar with them. Podcasts provide a means to "soft sell" your products or services without actually selling them anything.
- Repurpose the content into books or other content: Another way to monetize the podcasts is by consolidating the content into some kind of book (advice from top entrepreneurs) or blog articles, which, in turn, can be monetized via affiliates.
- Syndicate your show on YouTube: More and more podcasters are simulcasting their podcasts by recording in video as well. Even if you don't have a video, you can still create an MP4 file with cool graphics and start a YouTube channel. And then you can monetize your YouTube channel.
- Charge your guests: If your podcast is popular, you can actually charge guests to appear on your podcast.
- Charge for old episodes: You can create a paywall for your old episodes and sell them on a subscription or a one-time fee. This is a twist on a traditional subscription service.

Plan Your Monetization Strategy: Now It's Your Turn

Check off the box beside each task as you complete it.

- ❏ 1. Review your podcast goals: 10 minutes.

Revisit your podcast goals. Are they the same as Day 1, or do you need to modify them? These goals will help you plan your monetization strategy.

❏ 2. Prioritize your top three ways of monetizing: 30 minutes.

Now that your podcast goals are clear, review the various strategies mentioned above and pick your top three ways to monetize your podcast, depending on your specific situation. You don't have to focus on only one strategy. You can combine multiple strategies for an income stream.

❏ 3. Initiate a plan of action: 20 minutes.

Now, start putting a plan together. The subsequent chapters will guide you on specific steps for each strategy mentioned in the traditional ways to monetize.

Daily Standup

Did you complete today's tasks?

❏ Yes
❏ No

If no, what do you need to carry over to work on tomorrow?

_____

What did you learn about your business (or yourself) today that will serve you in the future?

_____

Day 27: Monetizing: Sell Your Own Products and Services

Your Podcast Is an Extension of Your Existing Business

Making More Money with What You Already Sell

Podcasts provide an excellent opportunity to pitch your existing products or services. Depending on the format of your podcast, you could talk about the benefits of your products or services or have the guests talk about your products.

After I published my first book, *The 60 Minute Startup*, I used to talk about the concepts in the book on my podcast. Other times, the guests themselves would bring that book up, and it almost became a testimonial without me having to pitch the book. In other words, my podcast provided me an opportunity to sell more of my books without a hard sell.

If you don't have a product or service to sell, don't worry. The next section is all about that.

> "I monetize my podcast by offering my own products and services, which you can see here: TheAdvisorCoach.com/Products. For example, if I do a podcast episode about LinkedIn marketing, I will recommend my LinkedIn product designed to help financial advisors get more clients."
> —James Pollard, Host, *The Financial Advisor Marketing Podcast*

> "I have started to tap into that via affiliate marketing and having guests on the show that are my ideal client and being able to sell them into my programs and products behind the scenes further delivering value. Soon I will be featuring my new digital courses on the show. Which will drive additional revenue via the podcast and all the dark horses that listen to it. I mean after all, *The Dark Horse Entrepreneur* is a podcast focused on those dark horses that want to start, restart, kickstart or start stepping up their business game . . . all with a personal development twist because we know that if your mindset is jacked your biz-set will get jacked."
> —Tracy Brinkmann, Host, *The Dark Horse Entrepreneur*

Create New Products or Services to Sell

The podcasts actually make it relatively easy to create new products or services. Your show will give you certain credibility and make you an expert in your selected domain. You can use this credibility and expertise to create your own products or services.

For example, let's say your show is about social media advertising. You can actually create a service by offering social media advertising expertise to your listeners. If you don't believe that you are an expert or don't have time, you can offer your guests services with an added markup.

Coaching and Consulting

Coaching and consulting are among the easier services you can offer with your podcast. Over a period of time, the conversations you may have with your guests or the research you may be doing about your topics will make you an expert without you knowing.

The expertise translates easily into becoming a coach or offering some part-time consulting services. I know of many podcast hosts who use their podcasting gig to start their coaching or consulting careers in many different fields.

Selling Your Products and Services: Now It's Your Turn

Check off the box beside each task as you complete it.

❏ 1. Make a decision on existing or new products: 10 minutes.

What will it be? New products? Existing products? Assess your situation, list your inventory down, and prioritize.

❏ 2. Put a plan together to pitch your products: 30 minutes.

For your selected products, highlight key benefits. Plan how you'll weave the benefits into your podcast episodes. Will you provide some promotions for them? Plan the integration of your products into your show.

❏ 3. Research ways to extend into coaching and consulting: 20 minutes.

Even if you have existing products, spend some time thinking about extending your show to create new opportunities like coaching and consulting. It may be a relatively simple step.

Daily Standup

Did you complete today's tasks?

❏ Yes
❏ No

If no, what do you need to carry over to work on tomorrow?

_____

What did you learn about your business (or yourself) today that will serve you in the future?

_____

Day 28: Monetizing: Affiliates and Commission

Recurring Small Percentages Can Add Up to Big Bucks

What is Affiliate Marketing?

There are companies that offer a commission from the sale of a product if the buyer came to them by clicking on a link from your website. As an affiliate of their business, you would sign an agreement before getting your own personal affiliate links for their products. The percentage of commissions varies anywhere from single digits to as high as 50 percent for some products.

The most popular of these affiliate merchants is Amazon, with a very well-established affiliate program, even though it pays in the low single digits. The merchant then allows you to generate your personalized affiliate links that you embed in articles on your site. And the cool thing is that you'll get a commission on any products (and not just the product listed on your site) that the buyer clicks on the merchant site within a given time.

Your job will be to identify the most relevant merchants for your podcast and include their links in articles and podcast transcripts on your site. So where would you find these merchants? Let's find out in the next section.

> "I monetize my podcast by reaching out to brands for affiliate links to feature, and you want to feature ads that may be relevant to your listeners (i.e., mine are figure skaters, so I featured athlete gut health tests and medicine balls), or if you sign with a podcast company, they also help you get ads."
>
> —Polina Edmunds, Host, *Bleav in Figure Skating*

> "I interview brilliant women (mostly entrepreneurs and venture capitalists) about their audacious vision for the world. Our listeners tend to say our podcast is incredibly inspiring, insightful, and motivates them to do more and be more. I publish every other week, and we use the podcast as a way of accessing high net-worth women that we would like to work with. Two of the women we interviewed have become advisors for our business, and others have facilitated introductions on our behalf."
>
> Melissa Kiguwa, Host, *Grit & Grace*

Primary Affiliate Sources

First, let's talk about established affiliate programs that you can tap into.

- Amazon Affiliate Program: **Very established program.**
- ShareASale: **A massive network with 3,900+ merchants and one million-plus affiliates with a nineteen-plus-year-old program.**
- CJAffiliate (formerly Commission Junction) is a twenty-year-old network with some big brands like Overstock, Lowes, Priceline, Office Depot, et cetera.

- Impact: Partnership relationship management (PRM) that connects brands to publishers.
- Clickbank: A B2C affiliate network well suited for health, wellness, food, nutrition, et cetera.

Secondary Affiliate Sources

Apart from the ones mentioned above, there are hundreds of other affiliate programs you can tap into.

- Rakuten Linkshare
- Lemonads

Your own podcast host might have an affiliate program. For example, Buzzsprout has an established affiliate program.

Your web hosting platform, like HostGator or GoDaddy.

Many course platforms like Teachable have affiliate programs.

SEO tools and email programs like SEMRush, GetResponse, et cetera.

For a much more exhaustive list, log into the member area on RameshDontha.com.

Affiliate Marketing: Now It's Your Turn

Check off the box beside each task as you complete it.

❏ 1. Identify relevant products and services: 20 minutes.

Do some basic research on relevant products or services for your podcast. If your podcast is about search engine optimization (SEO), write down products like web hosting, SEO tools, and email marketing tools, for example.

❏ 2. Sign up for affiliate programs: 30 minutes.

Most of the time, the merchant will have information about affiliate programs on their website. Otherwise, search on one of the affiliate networks mentioned above.

❑ 3. Get a couple of initial affiliate links: 10 minutes.

Some merchants approve your application right away, and others take time to approve. Once you are approved, create a couple of affiliate links and embed them in one of your podcast articles.

Daily Standup

Did you complete today's tasks?

❑ Yes
❑ No

If no, what do you need to carry over to work on tomorrow?

What did you learn about your business (or yourself) today that will serve you in the future?

Day 29: Monetization: Sponsorship and Advertising

Downloads + Quality = Sponsorships and Advertisers

Advertising

Podcast advertising is a growing industry. Just like in any media advertising, there are a growing number of middlemen who have been focusing on aggregating advertisers and matching them with relevant podcasts. A list of them can be found later in this chapter.

Advertisers are interested in the number of downloads and number of subscribers. Even though many podcast hosts look to advertising as their primary means of monetizing, it is fairly difficult to get advertisers for new podcasts, especially with unknown hosts. I do not want to discourage you from going down this path, but there is a lot less ROI for new podcasts.

There are multiple types of ads you can insert into an episode. The most common insertion points are pre-roll, mid-roll, and end. Pre-roll ads are inserted at the beginning, mid-roll ads are inserted midway through, and some ads can be inserted at the end as well. Ads are sold on a CPM (cost per thousand downloads) basis. Many of the podcast hosting platforms like Libsyn and Podbean offer means to insert ads into episodes.

> "Monetizing the show is a goal, and I've had two sponsors, but they weren't the right fit for my audience. Because my episodes are evergreen, I'm looking for the right sponsor who sees the value in a long-term relationship."
>
> —Jennifer Fink, Host, *Fading Memories*

> "Collaboration with guests and team members led to cross-promotional marketing efforts, and after a few seasons, we felt confident enough to start approaching sponsors. Being a podcast focused on social good bolstered the support of sponsors. These sponsorships, which give them the option to sponsor one, three, or an entire season of episodes, have since allowed for us to focus more on marketing efforts and is adding to our current rapid growth in audience."
>
> —Aalia Lanius, Host, *UNSUGARCOATED with Aalia*

Sponsorship

Sponsorship, in contrast to advertising, allows you to select the organizations you want to promote on your show. Unlike advertisers, sponsors may be willing to pay upfront without a large number of downloads if they like your content. Sponsors are more interested in the quality of your listeners than the quantity.

You can negotiate terms on a per-episode basis, per series, or per a fixed time period. Just like advertising, though, you'd be inserting sponsors' messages pre-roll, mid-roll, or at the end. Additionally, you may negotiate to include their logos and brand elements in show notes and on your website.

Unlike advertising, there are no networks to find sponsors. It depends on your networking skills to find the right sponsors that believe in you.

Resources for Sponsors and Advertisers

Here are some advertising networks for your podcast.

- Midroll
- Authentic
- Audacy
- AudioGO

Your own podcast hosting company, like Buzzsprout, Castos, Podbean, Captivate, et cetera.

Sponsorship and Advertising: Now It's Your Turn

Check off the box beside each task as you complete it.

❑ 1. Research potential advertisers: 20 minutes.

Go through the list of advertisers on the ad networks mentioned above and see which ones fit your podcast profile.

❑ 2. Shortlist of potential sponsors: 20 minutes.

Write down a list of potential sponsors from your network.

❑ 3. Reach out to sponsors and advertisers: 20 minutes.

Now take the time to reach out to the sponsors and advertisers.

Daily Standup

Did you complete today's tasks?

❑ Yes
❑ No

If no, what do you need to carry over to work on tomorrow?

_____

What did you learn about your business (or yourself) today that will serve you in the future?

_____

Day 30: Crowdfunding, Premium Content

Listeners Will Pay for Quality Content

Listener Donations

Podcasts have a one-to-one feeling. As a host, you'd believe that there is that one "ideal" listener, and that listener also feels that you are talking to him/her. Given that, those ideal listeners will be willing to support you. The question is how to get these listeners to pay?

Fear not! Patreon to the rescue. Patreon has been around for a long time, and the platform's primary purpose is to connect creators and their audiences in a meaningful way. In the process, Patreon has enabled payments and provided a way to create different tiers as well.

Another way to get listener donations is to have a tip jar or equivalent on your website. PayPal, Zelle, and Venmo provide easy ways to integrate the payments on your website. You'll be surprised how many people will be willing to "tip" for free content.

> "DO NOT be boring. Though some may want to hear only the facts on your subject, the majority of the listeners have a short attention span. Learn to be spontaneous and not robotic. Give them what they

want with some entertainment. Finally, be patient as it may take years for your show to catch on because most of the successful podcasters have been at it for a number of years before they see success."

—Frederick Penney, Host, *Radio Law Talk*

"The biggest problem for first-time podcasters is imposter syndrome. Not even tech. It's just purely the idea that you need to get out of your comfort zone. I was a part of Seth Godin's workshop for podcasters, and that definitely helped improve the confidence."

Natalie Luneva, Host, *SaaS Boss*

Premium Content

You can create premium content in many different ways. Here are some examples.

- "Special" interviews
- Behind-the-scenes content
- Ad-free RSS feed
- Early access to episodes
- Q&A with the hosts

The strategy to make this work is to have good-quality free content to attract many listeners and focus on some killer episodes only for premium subscribers. One cautionary note is not to alienate free listeners, and you can do that by making sure that the additional content is an extension of free content.

Merchandising

Another way to have listeners pay you is by creating merchandise. Even though it involves some upfront cost for you to create the merchandise based on your show, it may be worth it if you have a unique angle and dedicated audience.

There are many print-on-demand sites where you can upload your graphic/designs and have the audience order directly from them. Here are the most popular ones.

- Printful
- Redbubble
- Zazzle
- Spreadshirt
- CafePress

Crowdfunding: Now It's Your Turn

Check off the box beside each task as you complete it.

❏ 1. Patreon account: 20 minutes.

If you want to go this route, create a Patreon account and add info about your podcast. Don't forget to create at least two tiers.

❏ 2. Premium content: 20 minutes.

Brainstorm on different ways to separate "premium content" along the lines mentioned above.

❏ 3. Other monetization options: 20 minutes.

On Day 26, I mentioned many other monetization options. Check them out once again to see if anything suits your show.

Daily Standup

Did you complete today's tasks?

❑ Yes
❑ No

If no, what do you need to carry over to work on tomorrow?

What did you learn about your business (or yourself) today that will serve you in the future?

Your Podcast Is Live
. . . Now What?

Wow, you've made it! You have just crossed the thirty-day marathon finish line after sprinting daily for sixty minutes. That is an accomplishment worth celebrating!

If you have just been reading the book first to understand the concepts before actually implementing them, congratulations are still in order for going over all the key concepts of the agile podcaster's way to start a podcast in thirty days or fewer and make real progress toward monetization.

If you have implemented the strategies mentioned in the book for the past month or so, you are most likely looking at one of the following three scenarios.

- **Scenario One**: You've launched your podcast and now have at least one direct path to make money from your show.
- **Scenario Two**: You launched your podcast and are still thinking about the best way to monetize.
- **Scenario Three**: You were unable to launch your podcast because of some unexpected roadblocks.

For all three scenarios, the journey is far from over. There are multiple turns you can (and will) take to continue on your current path. If scenario one is your story, you have made a great accomplishment.

You've overcome the most significant hurdle any "podcastpreneur" will face—getting that first episode published. Now your task is to make money from your podcast.

For scenario two, your next key task is to choose one or more ways to monetize your content. You have a real podcast, and that matters. *A lot*. Now make your initial investment of time, effort, and money worth it. Get everything you can from this podcast. I recommend rereading the monetization chapters, then note which activity seems like it's the faster way to make money. Start there!

If scenario three best describes your situation, then you, my friend, have the easiest problem to fix. If you were unable to launch your podcast, more than likely the cause was inadequate resources. Not enough time, no money, not the right skills. So let's first figure out the underlying cause. Can you fix that issue? If you didn't have enough time, do you think you can find time in your schedule to go back and do the activities you skipped? If money was the problem, are you in a position to shore up finances without going broke? If it's a missing skill, can you outsource or otherwise get help from family or friends to fill that gap? Perhaps someone you know has the skills your show needs—and an interest in podcasting. In that case, the simple solution that changes everything for you might be a cohost partnership.

This journey is exhilarating and exhausting. Many veteran podcasters will admit that the happiest and most depressing times of their lives have been during their journey. Sometimes both at the same time. My effort with this book is to give this journey more balance. I want you to focus only on activities that lead to monetization. After all, that is the only goal of a business, as Peter Drucker said. Rid yourself of all other noises. Keep asking yourself what innovation and what marketing you could do to convert your content into cash.

And if you need help, I am just one website away at www.the60minutestartup.com. If you have any questions, feel free to reach out to me at Contact@the60minutestartup.com. I promise to respond to you as soon as I can. (You can connect with me on LinkedIn as well.)

Thank you for reading *The 60-Minute Podcast Startup*, and congratulations on reaching this far. Your exhilarating journey has only begun!

Acknowledgements

Writing a book is an all-encompassing, time-consuming project but extremely gratifying. I am thankful to so many people who have helped shape my thoughts and have inspired me with their accomplishments. *The 60-Minute Podcast Startup* is now a reality thanks to you.

I want to thank all of the featured podcast hosts in this book who opened up and shared their entrepreneurial journey with me. This book wouldn't have come alive without their personal stories.

I want to express my gratitude to Joshua Lisec, who has been the "X factor" in this and previous Agile Entrepreneurship Series book projects. He was my writing partner, guide, and editor of this book. I also want to acknowledge Joshua for coming up with the original book title *The 60-Minute Startup*.

I want to thank Ali Saif, who has been my technology partner with www.RameshDontha.com and has helped me launch The Agile Entrepreneurship Podcast in record time, ultimately inspiring this and previous books.

I want to thank all my friends at Folsom Rotary for encouraging me in all my endeavors.

Finally and most importantly, I want to thank my family, starting with my wife Sunanda for accepting all my crazy ideas, encouraging me, and allowing me to explore the uncharted territories. And I immensely thank our two daughters, Megha and Nidhi, for helping me become a better parent, for constantly motivating me, and for providing specific feedback on the book cover.

About the Author

Ramesh K. Dontha is a serial entrepreneur, bestselling author, and host of iTunes Top 500 podcast *The Agile Entrepreneurship Podcast*, iTunes Top 250 podcast *The Data Transformers*, and BrightTALK top-ranking web series *AI — The Future of Business*.

Ramesh is author of The 60-Minute Startup Series: Readers Favorite award-winner *The 60-Minute Startup*, #1 international bestseller *The 60-Minute Tech Startup*, and podcast launch guide *The 60-Minute Podcast Startup*. As a Fortune 100 consultant, Ramesh leveraged Agile principles to optimize technology systems. He then applied Agile to entrepreneurship, starting, growing, and selling multiple businesses. Now Ramesh teaches aspiring entrepreneurs how to get paying customers for theirs. Start your business and podcast in sixty minutes a day at www.The60MinuteStartup.com.

www.ingramcontent.com/pod-product-compliance
Lightning Source LLC
Chambersburg PA
CBHW020155200326
41521CB00006B/379